I Drink Your Milkshake
Fracking and the Indiscriminate Theft of America's Natural Resources

Contents

1 Environmental impact of hydraulic fracturing in the United States 1

   1.1 Air quality and methane emissions 2

   1.2 Water issues 2

      1.2.1 2015 EPA Report on Spills 2

      1.2.2 Water usage 3

      1.2.3 Injected fluid 3

      1.2.4 Groundwater contamination 4

      1.2.5 Flowback 5

      1.2.6 Surface water contamination 5

      1.2.7 Radioactivity 6

   1.3 Seismicity 8

      1.3.1 Induced seismicity from hydraulic fracturing 8

      1.3.2 Induced seismicity from water disposal wells 8

   1.4 Abandoned wells 9

   1.5 Health effects 10

      1.5.1 Worker health 11

      1.5.2 Research and lobbying 11

   1.6 Built Environment/Infrastructure 12

   1.7 See also 13

   1.8 References 13

   1.9 Further reading 20

   1.10 External links 21

2 Hydraulic fracturing 23

   2.1 Geology 23

      2.1.1 Mechanics 23

      2.1.2 Veins 24

      2.1.3 Dikes 24

   2.2 History 25

      2.2.1 Precursors 25

2.2.2   Oil and gas wells   25

2.2.3   Massive fracturing   26

2.2.4   Shales   26

2.3   Process   28

2.3.1   Method   28

2.3.2   Well types   29

2.3.3   Fracturing fluids   29

2.3.4   Fracture monitoring   32

2.3.5   Horizontal completions   33

2.4   Uses   34

2.5   Economic effects   35

2.6   Public debate   36

2.6.1   Politics and public policy   36

2.6.2   Documentary films   37

2.6.3   Research issues   37

2.7   Health risks   37

2.8   Environmental impacts   38

2.9   Regulations   39

2.10   See also   39

2.11   References   40

2.12   External links   46

2.13   Further reading   47

3   Regulation of hydraulic fracturing   50

3.1   Approaches   50

3.1.1   Risk-based approach   50

3.1.2   Precaution-based approach   51

3.2   Framing of the debate   51

3.2.1   "Learning-by-doing" and the displacement of ethics   51

3.2.2   Variations in risk-assessment of environmental impacts of hydraulic fracturing   52

3.3   Institutional discourse and the public   52

3.4   See also   53

3.5   References   53

3.6   External links   54

4   List of additives for hydraulic fracturing   55

4.1   Additives used in the United States   55

4.2   See also   55

4.3   References   55

5   Baldwin Hills Dam disaster   **56**

   5.1   Significance and diagnoses of the failure 56

   5.2   Coverage 59

   5.3   See also 59

   5.4   Notes 59

   5.5   References 60

   5.6   External links 61

6   Canadian Association of Petroleum Producers   **62**

   6.1   History 62

       6.1.1   CAPP Hall of Fame 63

   6.2   Advocacy for oil industry 63

       6.2.1   Advocacy for fracking 63

       6.2.2   Criticisms and concerns 63

   6.3   Advocacy for Crude Oil Exports via Canada's West Coast 63

       6.3.1   Criticisms and concerns 64

   6.4   Advocacy for Keystone XL Pipelines expansion 64

       6.4.1   Criticisms and concerns 64

   6.5   Advocacy regarding GHG emissions 64

       6.5.1   Criticisms and concerns 64

   6.6   CAPP initiative 2011: Oil and gas, industry, provincial regulators collaborate on strategies to shape public perception of fracking, water use and shale gas development 65

       6.6.1   Criticisms and concerns 65

   6.7   Selected CAPP publications 65

   6.8   See also 66

   6.9   Notes 66

   6.10 References 66

   6.11 Further reading 67

   6.12 External links 67

7   Canol shale play   **68**

   7.1   References 68

8   Chevron CRUSH   **70**

   8.1   History 70

   8.2   Process 70

   8.3   Isolation of groundwater 70

   8.4   See also 71

   8.5   References 71

9 Environmental impact of hydraulic fracturing   **73**

  9.1 Air emissions 74

      9.1.1 Climate change 74

  9.2 Water consumption 75

  9.3 Water contamination 75

      9.3.1 Injected fluid 76

      9.3.2 Flowback 76

      9.3.3 Surface spills 76

      9.3.4 Methane 77

  9.4 Radionuclides 77

  9.5 Land usage 78

  9.6 Seismicity 78

      9.6.1 Induced seismicity from hydraulic fracturing 78

      9.6.2 Induced seismicity from water disposal wells 78

  9.7 Noise 79

  9.8 Safety issues 79

  9.9 Health risks 79

  9.10 Policy and science 80

  9.11 See also 80

  9.12 References 80

  9.13 Bibliography 86

10 Exemptions for hydraulic fracturing under United States federal law   **87**

  10.1 Hydraulic fracturing: background 87

  10.2 Clean Water Act 87

  10.3 Safe Drinking Water Act 88

  10.4 National Environmental Policy Act 89

  10.5 Resource Conservation and Recovery Act 90

  10.6 Emergency Planning and Community Right-to-Know Act 90

  10.7 Comprehensive Environmental Response, Compensation, and Liability Act (Superfund) 90

  10.8 Debates Surrounding Regulatory Exemptions 91

  10.9 References 91

11 ExxonMobil Electrofrac   **94**

  11.1 Technology 94

  11.2 See also 94

  11.3 References 94

12 Fracking hose   **95**

12.1 Hose materials 95

12.2 Hose selection 95

12.3 History 95

12.4 References 95

13 The FracTracker Alliance **96**

13.1 History 96

13.2 Current Initiatives 96

13.2.1 Mapping 96

13.3 References 97

13.4 See also 97

13.5 External Links 97

14 Fracturing Responsibility and Awareness of Chemicals Act **98**

14.1 Background 98

14.2 Current status 99

14.3 See also 99

14.4 References 99

14.5 External links 99

15 Hydraulic fracturing proppants **100**

15.1 Proppant permeability and mesh size 100

15.2 Proppant weight and strength 101

15.3 Proppant deposition and post-treatment behaviours 101

15.4 Proppant costs 102

15.5 Other components of fracturing fluids 102

15.6 See also 102

15.7 References 102

16 Hydro-slotted perforation **104**

16.1 Overview 104

16.2 General concepts 105

16.2.1 Benefits 106

16.3 Development 107

16.4 Patents 108

16.5 References 108

17 Mohamed Yousef Soliman **109**

17.1 Biography 109

17.2 Most recent peer-reviewed papers 109

17.3 References 110

18 Oil and Gas Commission   **111**

18.1 Overview 111

    18.1.1 Purpose 111

    18.1.2 Tools 112

    18.1.3 Compliance and enforcement information 112

18.2 Lawsuits 112

18.3 Criticism 112

18.4 See also 112

18.5 References 113

18.6 External links 113

19 Promised Land (2012 film)   **114**

19.1 Plot 114

19.2 Cast 115

19.3 Production 115

19.4 Fracking 116

19.5 Financing 116

19.6 Release 116

    19.6.1 Theatrical run 116

    19.6.2 Critical reception 117

19.7 Accolades 117

19.8 References 117

19.9 External links 118

20 Stephanie Hallowich, H/W, v. Range Resources Corporation   **119**

20.1 External links 119

21 Uses of radioactivity in oil and gas wells   **121**

21.1 Use of radioactive sources for logging 121

21.2 Radiotracers and markers 121

21.3 Regulation in the US 122

21.4 See also 122

21.5 References 122

21.6 Text and image sources, contributors, and licenses 125

    21.6.1 Text 125

    21.6.2 Images 127

    21.6.3 Content license 128

Chapter 1

Environmental impact of hydraulic fracturing in the United States

Main articles: Environmental impact of hydraulic fracturing and Hydraulic fracturing in the United States
Environmental impact of hydraulic fracturing in the United States has been an issue of public concern, and includes

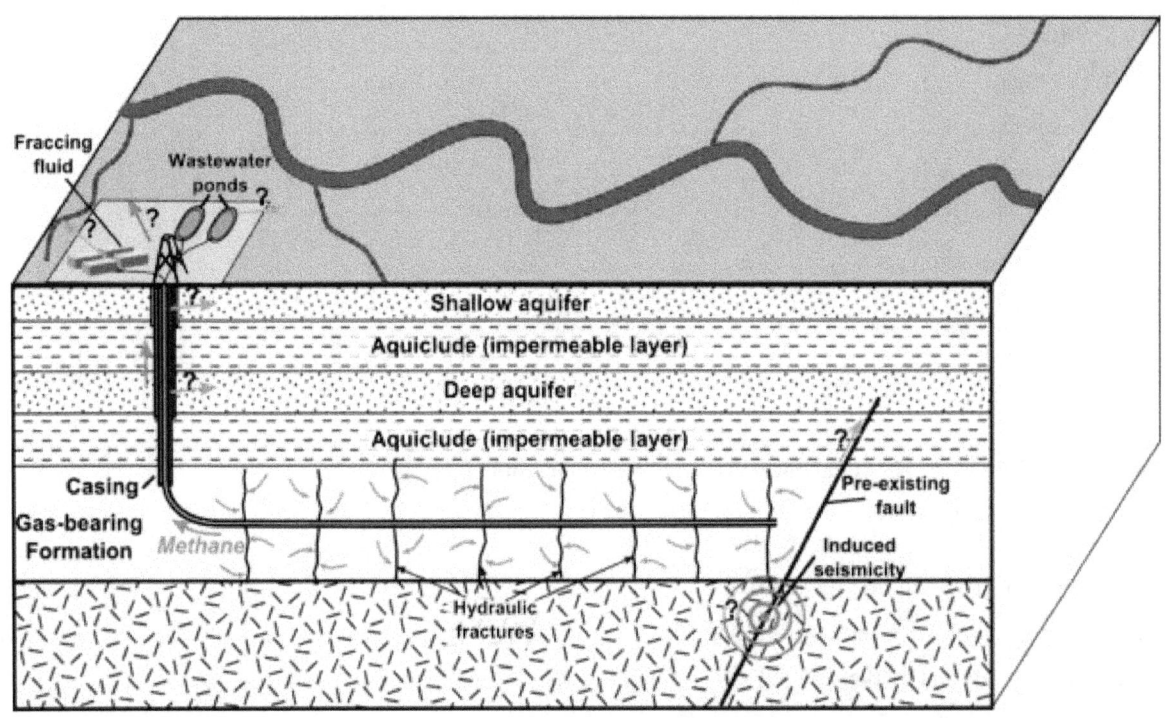

Schematic depiction of hydraulic fracturing for shale gas, showing potential environmental effects.

the potential contamination of ground and surface water, methane emissions,[1] air pollution, migration of gases and hydraulic fracturing chemicals and radionuclides to the surface, the potential mishandling of solid waste, drill cuttings, increased seismicity and associated effects on human and ecosystem health.[2][3] A number of instances with groundwater contamination have been documented,[4] however opponents of water safety regulation claim hydraulic fracturing has never caused any drinking water contamination.[5]

As early as 1987, researchers at the United States Environmental Protection Agency (EPA) concluded that hydraulic fracturing can contaminate and has contaminated groundwater. According to former EPA employees, evidence of the negative environmental impact of fracking was systematically removed from congressional reports to support the energy industry under the direction of the Office of Legal Counsel during the Reagan administration.[6] With the growth of hydraulic fracturing in the United States, "public exposure to the many chemicals involved in energy development is expected to increase over the next few years, with uncertain consequences" per science writer Valerie Brown in 2007.[3]

1.1 Air quality and methane emissions

Methane emissions from wells raise global warming concerns. There is a 2,500 square-mile methane plume hovering over the Four Corners area of the western US.[7] The magnitude of the plume was such that NASA researcher Christian Frankenberg reported to the press that, "We couldn't be sure that the signal was real."[8] According to NASA: "The study's lead author, Eric Kort of the University of Michigan, Ann Arbor, noted the study period predates the widespread use of hydraulic fracturing, known as fracking, near the hot spot. This indicates the methane emissions should not be attributed to fracking but instead to leaks in natural gas production and processing equipment in New Mexico's San Juan Basin, which is the most active coalbed methane production area in the country."[9]

Other concerns are related to emissions from the hydraulic fracturing chemicals and equipment such as volatile organic compound (VOC) and ozone. In 2008, ozone concentrations in ambient air near drilling sites in Sublette County, Wyoming were frequently above the National Ambient Air Quality Standards (NAAQS) of 75 ppb[10] and have been recorded as high as 125 ppb. In DISH, Texas, elevated levels of disulfides, benzene, xylenes and naphthalene have been detected in the air, emitted from compressor stations.[11] In Garfield County, Colorado, an area with a high concentration of drilling rigs, VOC emissions increased 30% between 2004 and 2006.[3]

Researchers from the University of Michigan analyzed the emissions produced from the hydraulic fracturing equipment at the Marcellus Shale and Eagle Ford Shale plays, and concluded that hydraulic pumps accounted for about 83% of the total emissions in the hydraulic fracturing fleet. NOx emission ranged between 3,600–5,600 lb/job, HC 232–289 lb/job, CO 859–1416 lb/job, and PM 184–310 lb/job. If the fuel efficiencies of the hydraulic fracturing pumps are improved, the emissions can be reduced.[12]

On April 17, 2012, the EPA issued cost-effective regulations, required by the Clean Air Act, which include the first federal air standards for natural gas wells that are hydraulically fractured.[13] The final rules are expected to yield a nearly 95% reduction in VOC emissions from more than 11,000 new hydraulically fractured gas wells each year. This reduction would be accomplished primarily through capturing natural gas that escapes into the air, and making it available for sale. The rules also will reduce air toxics, which are known or suspected of causing cancer and other serious health effects, and emissions of methane, a potent greenhouse gas.[13]

Stanford University in the American Journal of Science, looked at 200 studies and claimed the United States Environmental Protection Agency has been underestimating US methane emissions.[14] A survey of hydraulic fracturing sites in Pennsylvania revealed drilling operations releasing plumes of methane 100 to 1,000 times the rate the EPA expects from that stage of drilling.[15]

1.2 Water issues

1.2.1 2015 EPA Report on Spills

In May 2015, the EPA released a report reviewing the spill data from various state and industry sources for data about spills related to hydraulic fracturing.[16] Of the total reports reviewed in the study 1% (457) were determined to be related to hydraulic fracturing, while 66% were unrelated and 33% had insufficient data reported to determine if the spill was associated to hydraulic fracturing. In 324 incidents the spilled fluids were reported to reach categorized environmental receptors: Surface Water 67%, Soil 64%, and Ground Water 48%.

Other key spill figures from the report:

- Median spill volume 730 gallons

- The highest number and volume of spills from flowback/produced water

- Total fluid spilled 2,300,000 gallons

- Fluid recovered 480,000 gallons

- Fluid unrecovered 1,600,000 gallons

- Fluid unknown (recovery not reported) 250,000 gallons

- The largest numbers of spills were caused by human error 150 (33%); while the largest volume of spilled fluids was from failure of containers 1,500,000 gal (64%).

This report was cited in the full hydraulic fracturing water report now open for peer review "Assessment of the Potential Impacts of Hydraulic Fracturing for Oil and Gas on Drinking Water Resources"[17] though not directly addressed in the contents of the EPA spill data report it is interesting to note highlights several times where associations to hydraulic fracturing, causes of spills, and response to spills were unknown or indeterminate because of missing or unreported data. This highlights the need for more complete reporting and standardization of reporting for improved tracking to better guide implementation of environmental safety practices particularly where the impact is likely to impact key health determents like water quality.

1.2.2 Water usage

Hydraulic fracturing uses between 1.2 and 3.5 million US gallons (4,500 and 13,200 m^3) of water per well, with large projects using up to 5 million US gallons (19,000 m^3). Additional water is used when wells are refractured.[18][19] An average well requires 3 to 8 million US gallons (11,000 to 30,000 m^3) of water over its lifetime.[19][20][21][22] Back in 2008 and 2009 at the beginning of the shale boom in Pennsylvania, hydraulic fracturing accounted for 650 million US gallons per year (2,500,000 m^3/a) (less than 0.8%) of annual water use in the area overlying the Marcellus Shale.[20][21][23] The annual number of well permits, however, increased by a factor of five[24] and the number of well starts increased by a factor of over 17 from 2008 to 2011.[25]

According to Environment America, a federation of state-based, citizen-funded environmental advocacy organizations, there are concerns for farmers competing with oil and gas for water.[26] A report by Ceres questions whether the growth of hydraulic fracturing is sustainable in Texas and Colorado as 92% of Colorado wells were in extremely high water stress regions (that means regions where more than 80% of the available water is already allocated for agricultural, industrial and municipal water use) and 51% percent of the Texas wells were in high or extremely high water stress regions.[27] In Barnhart, Texas the aquifer supplying the local community ran dry because of intensive water utilization for hydraulic fracturing.[28] In 2013, the Railroad Commission of Texas adopted new hydraulic fracturing water recycling rules intended to encourage Texas hydraulic fracturing operators to conserve water used in the hydraulic fracturing process.[29]

Consequences for agriculture have already been observed in North America. In some regions of the US that are vulnerable to droughts, farmers are now competing with fracking industrials for the use of water resources.[30] In the Barnett Shale region, in Texas and New Mexico, drinking water wells have dried up due to fracking's withdrawal of water, and water has been taken from an aquifer used for residential and agricultural use.[30] Farmers have seen their wells go dry in Texas and New Mexico as a result of fracking's pressure on water resources, for instance in Carlsbad, New Mexico.[30] Agricultural communities have already seen water prices rising because of that problem. In the North Water Conservation District in Colorado was organized an auction to allocate water and the prices rose from $22/acre-foot in 2010 to $28 in the beginning of 2012.[30]

1.2.3 Injected fluid

Hydraulic fracturing fluids include proppants, radionuclide tracers, and other chemicals. While many are common and generally harmless, some additives used in the United States are known carcinogens.[2] Out of 2,500 hydraulic fracturing products, more than 650 contained known or possible human carcinogens regulated under the Safe Drinking Water Act or listed as hazardous air pollutants".[2] Between 2005 and 2009, 279 products had at least one component listed as "proprietary" or "trade secret" on their Occupational Safety and Health Administration (OSHA) required safety data sheet (SDS). In many instances, companies who bought products off the shelf did not know the ingredients.[2] Without knowing the identity of the proprietary components, regulators cannot test for their presence. This prevents government regulators from establishing baseline levels of the substances prior to hydraulic fracturing and documenting changes in these levels, thereby making it more difficult to prove that hydraulic fracturing is contaminating the environment with these substances.[31]

The Ground Water Protection Council launched FracFocus.org, an online voluntary disclosure database for hydraulic fracturing fluids funded by oil and gas trade groups and the United States Department of Energy (DOE). The site has been met with some scepticism relating to proprietary information that is not included.[32][33] Some states have mandated fluid disclosure and incorporated FracFocus as the tool for disclosure.[34][35]

1.2.4   Groundwater contamination

In 2009, state regulators from across the country stated that they had seen no evidence of hydraulic fracturing contaminating water in their respective jurisdictions.[36] In May 2011 the EPA Administrator Lisa P. Jackson testified in a Senate Hearing Committee stating that the EPA had never made a definitive determination of contamination where the hydraulic fracturing process itself has contaminated water.[37] However, by 2013, Dr. Robin Ikeda, Deputy Director of Non-communicable Diseases, Injury and Environmental Health at the CDC testified to congress that EPA had documented contamination at several sites.[38]

Incidents of contamination

- As early as 1987, an EPA report was published that indicated fracture fluid invasion into James Parson's water well in Jackson County, West Virginia. The well, drilled by Kaiser Exploration and Mining Company, was found to have induced fractures that created a pathway to allow fracture fluid to contaminate the groundwater from which Mr. Parson's well was producing.[39] Directed by Congress, the EPA announced in March 2010 that it will examine claims of water pollution related to hydraulic fracturing.[40][41]

- In 2006, over 7 million cubic feet (200,000 m^3) of methane were released from a blown gas well in Clark, Wyoming and nearby groundwater was found to be contaminated with hydrocarbon compounds and benzene particularly.[42][43]

- An investigation was initiated after a Pennsylvania water well exploded on New Year's Day in 2009. The state investigation revealed that Cabot Oil & Gas Company "had allowed combustible gas to escape into the region's groundwater supplies."[44][45] Arsenic, barium, DEHP, glycol compounds, manganese, phenol, methane, and sodium were found in unacceptable levels in the wells.[46] In April 2010, the state of Pennsylvania banned Cabot Oil & Gas Corp. from further drilling in the entire state until it plugs wells believed to be the source of contamination of the drinking water of 14 homes in Dimock Township, Pennsylvania.[47] Cabot Oil & Gas was also required to financially compensate residents and provide alternative sources of water until mitigation systems were installed in affected wells.[46] The company denies, however, that any "of the issues in Dimock have anything to do with hydraulic fracturing".[48][49][50] In May 2012 the EPA reported that their most recent "set of sampling did not show levels of contaminants that would give the EPA reason to take further action." Methane was found only in one well.[51] Cabot has held that the methane was preexisting, but state regulators have cited chemical fingerprinting as proof that it was from Cabot's hydraulic fracturing activities.[52] The EPA plans to re-sample four wells where previous data by the company and the state showed levels of contaminants.[51]

- Complaints about water quality from residents near a gas field in Pavillion, Wyoming prompted an EPA groundwater investigation. An EPA draft report dated December 8, 2011 found that contaminants in surface water near pits indicated were a source of contamination, and by the time the report was issued the company had already started to remediate the pits.[53] The report also suggested that the groundwater contained "compounds likely associated with gas production practices, including hydraulic fracturing... Alternative explanations were carefully considered for individual sets of data. However, when considered together with other lines of evidence, the data indicates likely impact to ground water that can be explained by hydraulic fracturing."[53] The Agency for Toxic Substances and Disease Registry recommended that owners of tainted wells use alternate sources of water for drinking and cooking, and ventilation when showering. Encana is funding the alternate water supplies.[54] State and industry figures rejected the EPA's findings.[55] In 2012 the U.S. Geological Survey, tasked with further sampling of the EPA wells, tested one of two EPA monitoring wells near Pavillion (the other well the USGS considered unsuitable for collecting water samples) and found evidence of methane, ethane, diesel compounds and phenol,[56] In June 2013, the EPA announced that it was closing its investigation at Pavilion, and would not finish or seek peer review of its preliminary 2011 study. Further investigation will be done by the state of Wyoming.[57]

1.2.5 Flowback

Flowback is the portion of the injected fracturing fluid that flows back to the surface, along with oil, gas, and brine, when the well is produced. An estimated 90% of flowback in the United States is disposed of into deep EPA-licensed Class II disposal wells, with the remaining less than 10% reused, evaporated, used for irrigation, or discharged to surface streams under an NPDES permit. Of nine oil and gas-producing states studied in 2012, underground injection disposal was by far the predominant method in all but Pennsylvania where were only six active waste disposal wells.[58] In California, Virginia, and Ohio there have been instances of illegal dumping of flowback, a precursor to possible contamination of local ground and surface water reservoirs.[59] Discharging oil and gas produced water to surface streams without an NPDES permit is a federal crime.[60] Discharges through water treatment works must comply with the federal Clean Water Act and the terms of their NPDES permits, but the EPA noted that most water treatment works are not set up to treat flowback.[61]

In Pennsylvania, oil and gas produced water had for many years been accepted by licensed water treatment works for treatment and discharge, but the volume expanded greatly with the proliferation of Marcellus Shale wells after 2000. In 2010 the Pennsylvania Department of Environmental Protection (DEP) limited surface water discharges from new treatment plants to 250 mg/l chloride; the chloride limitation was designed to also limit other contaminants such as radium. Existing water treatment plants were "grandfathered," and still allowed higher discharge concentrations, but oil and gas operators were prohibited to send wastewater to the grandfathered treatment plants.[62]

One Duke University study reported that "Marcellus [Shale] wells produce significantly less wastewater per unit gas recovered (~35%) compared to conventional natural gas wells."[63] In Colorado the volume of wastewater discharged to surface streams increased from 2008 to 2011.[64]

1.2.6 Surface water contamination

Hydraulic fracturing can affect surface water quality either through accidental spills at the wellsite, or by discharge of the flowback through existing water treatment works. Directed by Congress, the EPA announced in March 2010 that it would examine claims of water pollution related to hydraulic fracturing.[40] Christopher Portier, director of the CDC's National Center for Environmental Health and the Agency for Toxic Substances and Disease Registry, argued that, in addition to the EPA's plans to investigate the impact of hydraulic fracturing on drinking water, additional studies should be carried out to determine whether wastewater from the wells can harm people or animals and vegetables they eat.[65] A group of US doctors called for a moratorium on hydraulic fracturing in populated areas until such studies had been done.[66][67]

However, others point out exclusions and exemptions for hydraulic fracturing under United States federal law. Exemptions were made in the Clean Water Act, as part of the Energy Policy Act of 2005, also known as the "Halliburton Loophole." These exemptions included stormwater runoff from gas and oil construction activities which includes "oil and gas exploration, production, process, or treatment operations and transmission facilities" as part of the definition of construction activities.[68] Amendments to the Safe Drinking Water Act involved the definition of underground injection. Underground injection related to hydraulic fracturing was exempted from the Clean Water Act, except if it uses diesel fuel.[69]

The growing of oil and natural gas drilling employing hydraulic fracturing technology is steady around different regions of the United States, but the maintenance of wastewater gathered after the drilling process containing hydraulic fracturing fluids is lagging behind.[70] In Pennsylvania, the DEP reported that the resources to properly regulate wastewater-handling facilities were unavailable, inspecting facilities every 20 years rather than every 2 years as called for by regulation.[70]

The quantity of wastewater and the unpreparedness of sewage plants to treat wastewater, is an issue in Pennsylvania.[71][72] The Associated Press has reported that starting in 2011, the DEP strongly resisted providing the AP and other news organizations with information about complaints related to drilling.[73] When waste brine is discharged to surface waters through conventional wastewater treatment plants, the bromide in the brine usually is not captured. Although not a health hazard by itself, in western Pennsylvania some downstream drinking water treatment plants using the surface water experienced increases in brominated trihalomethanes in 2009 and 2010. Trihalomethanes, undesirable byproducts of the chlorination process, form when the chlorine combines with dissolved organic matter in the source water, to form the trihalomethane chloroform. Bromine can substitute for some chlorine, forming brominated trihalomethanes. Because bromine has a higher atomic weight than chlorine, the partial conversion to brominated trihalomethanes increases the concentration by weight of total trihalomethanes.[74][75][76]

1.2.7 Radioactivity

See also: Radionuclides associated with hydraulic fracturing

Radioactivity associated with hydraulically fractured wells comes from two sources: naturally occurring radioactive material and radioactive tracers introduced into the wells. Flowback from oil and gas wells is usually disposed of deep underground in Class II injection wells, but in Pennsylvania, much of the wastewater from hydraulic fracturing operations is processed by public sewage treatment plants. Many sewage plants say that they are incapable of removing the radioactive components of this waste, which is often released into major rivers. Industry officials, though, claim that these levels are diluted enough that public health is not compromised.[71]

In 2011, the level of dissolved radium in hydraulic fracturing wastewater released upstream from drinking water intakes had been measured to be up to 18,035 pCi/L (667.3 Bq/l),[77] and the gross alpha level measured to be up to 40,880 pCi/L (1,513 Bq/l).[71][77] The New York Times reported that studies by the EPA and a confidential study by the drilling industry concluded that radioactivity in drilling waste cannot be fully diluted in rivers and other waterways.[78] A recent Duke University study sampled water downstream from a Pennsylvania wastewater treatment facility from 2010 through Fall 2012 and found the creek sediment contained levels of radium 200 times background levels.[79] The surface water had the same chemical signature as rocks in the Marcellus Shale formation. The facility denied processing Marcellus waste since 2011. In May 2013 the facility signed another agreement to not accept or discharge wastewater Marcellus Shale formations until it has installed technology to remove the radiation compounds, metals and salts.[80][81] According to the Duke researches the 'waste treatment solids/sludge' have exceeded U.S. regulations for radium disposal to soil.[80] The study by Duke University also found that radium has been "absorbed and accumulated on the sediments locally at the discharge".[80]

The New York Times noted that in 2011 the Pennsylvania DEP only made a "request — not a regulation" of gas companies to stop sending their flowback and waste water to public water treatment facilities.[82] However, the DEP gave oil and gas operators 30 days to voluntarily comply, and they all did.[62] Former Pennsylvania DEP Secretary John Hanger, who served under Gov. Ed Rendell, affirmed that municipal drinking water throughout the state is safe. "Every single drop that is coming out of the tap in Pennsylvania today meets the safe drinking water standard," Hanger said, but added that the environmentalists were accurate in stating that Pennsylvania water treatment plants were not equipped to treat hydraulic fracturing water.[83] Current Pennsylvania DEP Secretary Michael Krancer serving under Gov. Tom Corbett has said it is "total fiction" that untreated wastewater is being discharged into the state's waterways,[84] though it has been observed that Corbett received over a million dollars in gas industry contributions,[85] more than all his competitors combined, during his election campaign.[86] Unannounced inspections are not made by regulators: the companies report their own spills, and create their own remediation plans.[71] A recent review of the state-approved plans found them to appear to be in violation of the law.[71] Treatment plants are still not equipped to remove radioactive material and are not required to test for it.[71] Despite this, in 2009 the Ridgway Borough's public sewage treatment plant, in Elk County, PA, facility was sent wastewater containing radium and other types of radiation at 275–780 times the drinking-water standard. The water being released from the plant was not tested for radiation levels.[71] Part of the problem is that growth in waste produced by the industry has outpaced regulators and state resources.[71] It should be noted that "safe drinking water standards" have not yet been set for many of the substances known to be in hydrofracturing fluids or their radioactivity levels,[71] and their levels are not included in public drinking water quality reports.[87]

Tests conducted in Pennsylvania in 2009 found "no evidence of elevated radiation levels" in waterways.[88] At the time radiation concerns were not seen as a pressing issue.[88] In 2011 The New York Times reported radium in wastewater from natural gas wells is released into Pennsylvania rivers,[71][89] and compiled a map of these wells and their wastewater contamination levels,[77] and stated that some EPA reports were never made public.[78] The *Times'* reporting on the issue has come under some criticism.[90][91] A 2012 study examining a number of hydraulic fracturing sites in Pennsylvania and Virginia by Pennsylvania State University, found that water that flows back from gas wells after hydraulic fracturing contains high levels of radium.[92]

Before 2011, flowback in Pennsylvania was processed by public wastewater plants, which were not equipped to remove radioactive material and were not required to test for it. Industry officials, though, claim that these levels are diluted enough that public health is not compromised.[71][72] In 2010 the DEP limited surface water discharges from new treatment plants to 250 mg/l chloride. This limitation was designed to also limit other contaminants such as radium. Existing water treatment plants were allowed higher discharge concentrations. In April 2011, the DEP asked unconventional gas operators to voluntarily stop sending wastewater to the grandfathered treatment plants. The PADEP reported that the operators had complied.[62]

A 2013 Duke University study sampled water downstream from a Pennsylvania wastewater treatment facility from 2010 through 2012 and found that creek sediment contained levels of radium 200 times background levels.[79] The surface water had the same chemical signature as rocks in the Marcellus Shale formation along with high levels of chloride. The facility denied processing Marcellus waste after 2011. In May 2013 the facility signed another agreement to not accept or discharge Marcellus wastewater until it installed technology to remove the radioactive materials, metals and salts.[80][81]

A 2012 study by researchers from the National Renewable Energy Laboratory, University of Colorado, and Colorado State University reported a reduction in the percentage of flowback treated through surface water discharge in Pennsylvania from 2008 through 2011.[64] By late 2012, bromine concentrations had declined to previous levels in the Monongahela River, but remained high in the Allegheny.[93]

Naturally occurring radioactive materials

The New York Times has reported radiation in hydraulic fracturing wastewater released into rivers in Pennsylvania.[71] It collected data from more than 200 natural gas wells in Pennsylvania and has posted a map entitled *Toxic Contamination from Natural Gas Wells in Pennsylvania*. The *Times* stated "never-reported studies" by the United States Environmental Protection Agency and a "confidential study by the drilling industry" concluded that radioactivity in drilling waste cannot be fully diluted in rivers and other waterways.[78] Despite this, as of early 2011 federal and state regulators did not require sewage treatment plants that accept drilling waste (which is mostly water) to test for radioactivity. In Pennsylvania, where the drilling boom began in 2008, most drinking-water intake plants downstream from sewage treatment plants have not tested for radioactivity since before 2006.[71]

The New York Times reporting has been criticized[90] and one science writer has taken issue with one instance of the newspaper's presentation and explanation of its calculations regarding dilution,[94] charging that a lack of context made the article's analysis uninformative.[91]

According to a *Times* report in February 2011, wastewater at 116 of 179 deep gas wells in Pennsylvania "contained high levels of radiation," but its effect on public drinking water supplies is unknown because water suppliers are required to conduct tests of radiation "only sporadically".[95] The *New York Post* stated that the DEP reported that all samples it took from seven rivers in November and December 2010 "showed levels at or below the normal naturally occurring background levels of radioactivity", and "below the federal drinking water standard for Radium 226 and 228."[96] However, samples taken by the state from at least one river, (the Monongahela, a source of drinking water for parts of Pittsburgh), were taken upstream from the sewage treatment plants accepting drilling waste water.[97]

Radioactive tracers

Radioactive tracer isotopes are sometimes injected with hydraulic fracturing fluid to determine the injection profile and location of created fractures.[98] Sand containing gamma-emitting tracer isotopes is used to trace and measure fractures. A 1995 study found that radioactive tracers were used in over 15% of stimulated oil and gas wells.[99] In the United States, injection of radionuclides are licensed and regulated by the Nuclear Regulatory Commission (NRC).[100] According to the NRC, some of the most commonly used tracers include antimony-124, bromine-82, iodine-125, iodine-131, iridium-192, and scandium-46.[100] A 2003 publication by the International Atomic Energy Agency confirms the frequent use of most of the tracers above, and says that manganese-56, sodium-24, technetium-99m, silver-110m, argon-41, and xenon-133 are also used extensively because they are easily identified and measured.[101] According to a 2013 meeting of researchers who examined low (never exceeding drinking water standards) but persistent detections of iodine-131 in a stream used for Philadelphia drinking water: "Workshop participants concluded that the likely source of 131-I in Philadelphia's source waters is residual 131-I excreted from patients following medical treatments," but suggested that other potential sources also be studied, including hydraulic fracturing.[102]

1.3   Seismicity

Hydraulic fracturing routinely produces microseismic events much too small to be detected except by sensitive instruments. These microseismic events are often used to map the horizontal and vertical extent of the fracturing.[103] However, a 2012 US Geological Survey study reported that a "remarkable" increase in the rate of M ≥ 3 earthquakes in the US midcontinent "is currently in progress", having started in 2001 and culminating in a 6-fold increase over 20th century levels in 2011. The overall increase was tied to earthquake increases in a few specific areas: the Raton Basin of southern Colorado (site of coalbed methane activity), and gas-producing areas in central and southern Oklahoma, and central Arkansas.[104] While analysis suggested that the increase is "almost certainly man-made", the United States Geological Survey (USGS) noted: "USGS's studies suggest that the actual hydraulic fracturing process is only very rarely the direct cause of felt earthquakes." The increased earthquakes were said to be most likely caused by increased injection of gas-well wastewater into disposal wells.[105] The injection of waste water from oil and gas operations, including from hydraulic fracturing, into saltwater disposal wells may cause bigger low-magnitude tremors, being registered up to 3.3 (M_w).[106]

1.3.1   Induced seismicity from hydraulic fracturing

Hydraulic fracturing routinely triggers microseismic events too small to be detected except with sensitive instruments. However, according to the US Geological Survey: "Reports of hydraulic fracturing causing earthquakes large enough to be felt at the surface are extremely rare, with only three occurrences reported as of late 2012, in Great Britain, Oklahoma, and Canada."[107] Bill Ellsworth, a geoscientist with the U.S. Geological Survey, has said, however: "We don't see any connection between fracking and earthquakes of any concern to society."[108] The National Research Council (part of the National Academy of Sciences) has also observed that hydraulic fracturing, when used in shale gas recovery, does not pose a serious risk of causing earthquakes that can be felt.[109]

1.3.2   Induced seismicity from water disposal wells

Of greater concern are earthquakes associated with permitted Class II deep wastewater injection wells, many of which inject frac flowback and produced water from oil and gas wells. The USGS has reported earthquakes induced by disposal of produced water and hydraulic fracturing flowback into waste disposal wells in several location

In 2013, Researchers from Columbia University and the University of Oklahoma demonstrated that in the midwestern United States, some areas with increased human-induced seismicity are susceptible to additional earthquakes triggered by the seismic waves from remote earthquakes. They recommended increased seismic monitoring near fluid injection sites to determine which areas are vulnerable to remote triggering and when injection activity should be ceased.[110][111]

Geophysicist Cliff Frohlich researched seismic activity on the Barnett Shale in Texas from 2009 to 2011. Frohlich set up temporary seismographs on a 70 kilometer grid covering the Barnett Shale in Texas. The seismographs sensed and located earthquakes 1.5 magnitude and larger in the area. The seismographs revealed a spacial association between earthquakes and Class II injection wells, most of which were established to dispose of flowback and produced water from Barnett Shale wells, near Dallas-Fort Worth and Cleburne, Texas. Some of the earthquakes were greater than magnitude 3.0, and were felt by peole at the surface, and reported in the local news. Earthquakes were reported in areas where there had previously been no recorded earthquakes.[112] The study found that the great majority of Class II injection wells are not associated with earthquakes. Injection-induced earthquakes were strongly associated with wells injecting more than 150,000 barrels of water per month, and particularly after those wells had been injecting for more than a year. The majority of induced earthquakes occurred in Johnson County, which seemed more prone to induced earthquakes than other parts of the Barnett play.[113]

Earthquakes large enough to be felt by people have also been linked to some deep disposal wells that receive hydraulic fracturing flowback and produced water from hydraulically fractured wells. Flowback and brine from oil and gas wells are injected into EPA-regulated class II disposal wells. According to the EPA, approximately 144,000 such class II disposal wells in the US receive more than 2 billion US gallons (7.6 Gl) of wastewater each day.[114] To date, the strongest earthquakes triggered by underground waste injection were three quakes close to Richter magnitude 5 recorded in 1967 near a Colorado disposal well which received non-oilfield waste.[115]

According to the USGS only a small fraction of roughly 40,000 waste fluid disposal wells for oil and gas operations in the United States have induced earthquakes that are large enough to be of concern to the public.[116] Although the magnitudes of these quakes has been small, the USGS says that there is no guarantee that larger quakes will not occur.[117] In addition, the frequency of the quakes has been increasing. In 2009, there were 50 earthquakes greater than magnitude 3.0 in the area spanning Alabama and Montana, and there were 87 quakes in 2010. In 2011 there were 134 earthquakes in the same area, a sixfold increase over 20th century levels.[118] There are also concerns that quakes may damage underground gas, oil, and water lines and wells that were not designed to withstand earthquakes.[117][119]

The 2011 Oklahoma earthquake, the largest earthquake in Oklahoma history (most sources describe it as magnitude 5.7, although the US Geological Survey lists it as 5.6) has been linked by some researchers to decades-long injection of brine.[120] A 2015 study concluded that recent earthquakes in central Oklahoma, which includes 5.6 magnitude quake, were triggered by injection of produced water from conventional oil reservoirs in the Hunton Group, and are unrelated to hydraulic fracturing.[121]

Class II disposal wells receiving brine from Fayetteville Shale gas wells in Central Arkansas triggered hundreds of shallow earthquakes, the largest of which was magnitude 4.7, and caused damage. In April 2011, the Arkansas Oil and Gas Commission halted injection at two of the main disposal wells, and the earthquakes abated.[122]

Several earthquakes in 2011, including a 4.0 magnitude tremor on New Year's Eve that hit Youngstown, Ohio, are likely linked to a disposal of hydraulic fracturing wastewater,[110] according to seismologists at Columbia University.[123] By order of the Ohio Department of Natural Resources, the well had stopped injecting on December 30, 2011. The following day, after the 4.0 quake, Ohio governor John Kasich ordered an indefinite halt to injection in three additional deep disposal wells in the vicinity. The Department of Natural Resources proposed a number of tightened rules to its Class II injection regulations. The Department noted that there were 177 operational Class II disposal wells in the state, and that the Youngstown well was the first to produce recorded earthquakes since Ohio's Underground Injection Control program began in 1983.[124]

Since 2008, more than 50 earthquakes, up to a magnitude of 3.5, have occurred in the area of north Texas home to numerous Barnett Shale gas wells, an area that previously had no earthquakes. No injuries or serious damage from the earthquakes has been reported. A study of quakes near the Dallas-Fort Worth Airport 2008–2009, concluded that the quakes were triggered by disposal wells receiving brine from gas wells.[125]

A two-year study 2009–2011 by University of Texas researchers concluded that a number of earthquakes from Richter magnitude 1.5 to 2.5 in the Barnett Shale area of north Texas were linked to oilfield waste disposal into Class II injection wells. No quakes were linked to hydraulic fracturing itself.[126] Researchers noted that there are more than 50,000 Class II disposal wells in Texas receiving oilfield waste, yet only a few dozen are suspected of triggering earthquakes.[125]

On May 31, 2014, an earthquake registering at a magnitude of 3.4 occurred in Greeley, Colorado. The earthquake occurred near two hydraulic fracturing wastewater injection wells that are reportedly close to capacity. One waste injection well is 8,700 feet deep and 20 years old, while the other is 10,700 feet and just two years old. A research team from the University of Colorado Boulder have placed seismographs in the area to monitor further activity.[127][128]

1.4 Abandoned wells

Drilling for oil and gas has been going on in Pennsylvania since 1859, and there are an estimated 300,000 to 500,000 wells drilled before the state kept track of the wells, or required them to be properly plugged. The Pennsylvania Department of Environmental Protection (DEP) has a program to locate and plug old wells. A 2014 study examined 19 abandoned wells, 14 of which had never been plugged, and only one of which was known to the state. Methane leakage rates were measured, and extrapolations over all the expected orphaned wells in the state indicated that the old wells made up a significant source of methane.[129][130][131]

1.5   Health effects

There is worldwide concern over the possible adverse public health implications of hydraulic fracturing activity.[132] Intensive research is underway to ascertain whether there are impacts on a number of health conditions.[132]

Potential sources for ground and surface water exposure to toxins and toxicants (including endocrine disrupting hormones, heavy metals, minerals, radioactive substances, and salts) include 1) the drilling and fracturing phase; 2) improper treatment of wastewater, including spills during transport; and 3) failure of cement wall casings.

Many of the above contaminants have been associated with poor health outcomes, especially reproductive and developmental. Heavy metal and benzene/toluene exposure during pregnancy has been associated with miscarriage and stillbirths. Benzene and toluene have been associated with menstrual cycle disorders. Cancer, blood disorders, nervous system impairment, and respiratory issues have also been cited as potential complications of hydraulic fracturing fluid exposure.[133][134][135]

The 2014 EPA Executive summary describes evidence of drinking water contamination due to spills, inadequate casings, and other etiologies. Per this summary, frequency estimates range from one spill for every 100 wells in Colorado to between 0.4-12.2 spills for every 100 wells in Pennsylvania. Furthermore, "at least 3% of the wells (600 out of 23,000 wells) did not have cement across a portion of the casing installed through the protected ground water resource identified by well operators."[136]

While the health effects of water contamination, as well as air pollution and other potential health hazards due to hydraulic fracturing, is not well understood, studies report concerning findings. A 2014 retrospective cohort study of 124,842 births between 1996-2009 in rural Colorado reported statistically significant odds of congenital heart disease, including neural tube defects, with resident exposure to hydraulic fracturing.[134]

A 2015 study revealed lower birth weights and a higher incidence of small for gestational age comparing most to least exposed.[137]

A 2013 review focusing on Marcellus shale gas hydraulic fracturing and the New York City water supply stated, "Although potential benefits of Marcellus natural gas exploitation are large for transition to a clean energy economy, at present the regulatory framework in New York State is inadequate to prevent potentially irreversible threats to the local environment and New York City water supply. Major investments in state and federal regulatory enforcement will be required to avoid these environmental consequences, and a ban on drilling within the NYC water supply watersheds is appropriate, even if more highly regulated Marcellus gas production is eventually permitted elsewhere in New York State."[138]

Early in January 2012, Christopher Portier, director of the US CDC's National Center for Environmental Health and the Agency for Toxic Substances and Disease Registry, argued that, in addition to the EPA's plans to investigate the impact of fracking on drinking water, additional studies should be carried out to determine whether wastewater from the wells can harm people or animals and vegetables they eat.[65]

As of May 2012, the United States Institute of Medicine and United States National Research Council were preparing to review the potential human and environmental risks of hydraulic fracturing.[139][140]

In 2011 in Garfield County, Colorado, the U.S. Agency for Toxic Substances and Disease Registry collected air samples at 14 sites, including 8 oil and gas sites, 4 urban background sites, and 2 rural background sites. and detected carcinogens such as benzene, tetrachloroethene, and 1-4 dichlorobenzene at all the sites, both oil and gas sites, and background sites. Benzene was detected at 7 out of 8 oil and gas sites, in all 4 urban areas, and one out of the 2 rural background sites. The compound 1,4-dichlorobezene was detected in 3 out of 8 oil and gas sites, 3 out of 4 urban sites, and 1 out of 2 rural background sites. The benzene concentrations at one of the eight oil and gas sites was identified as cause for concern, because although it was within the acceptable range, it was near the upper limit of the range. The report concluded: "With the exception of the Brock site, these risk estimates do not appear to represent a significant theoretical cancer risk at any of the sites, nor does it appear that that the theoretical cancer risk is elevated at oil and gas development sites as compared to urban or rural background sites."[141][142]

In 2011, the EPA released new emissions guidelines stating that the old standards could have led to an unacceptably high risk of cancers for those living near drilling operations.[142]

1.5.1 Worker health

In 2013 the United States the Occupational Safety and Health Administration (OSHA) and the National Institute for Occupational Safety and Health (NIOSH) released a hazard alert based on data collected by NIOSH that "workers may be exposed to dust with high levels of respirable crystalline silica (silicon dioxide) during hydraulic fracturing."[143] NIOSH notified company representatives of these findings and provided reports with recommendations to control exposure to crystalline silica and recommend that all hydraulic fracturing sites evaluate their operations to determine the potential for worker exposure to crystalline silica and implement controls as necessary to protect workers.[144]

1.5.2 Research and lobbying

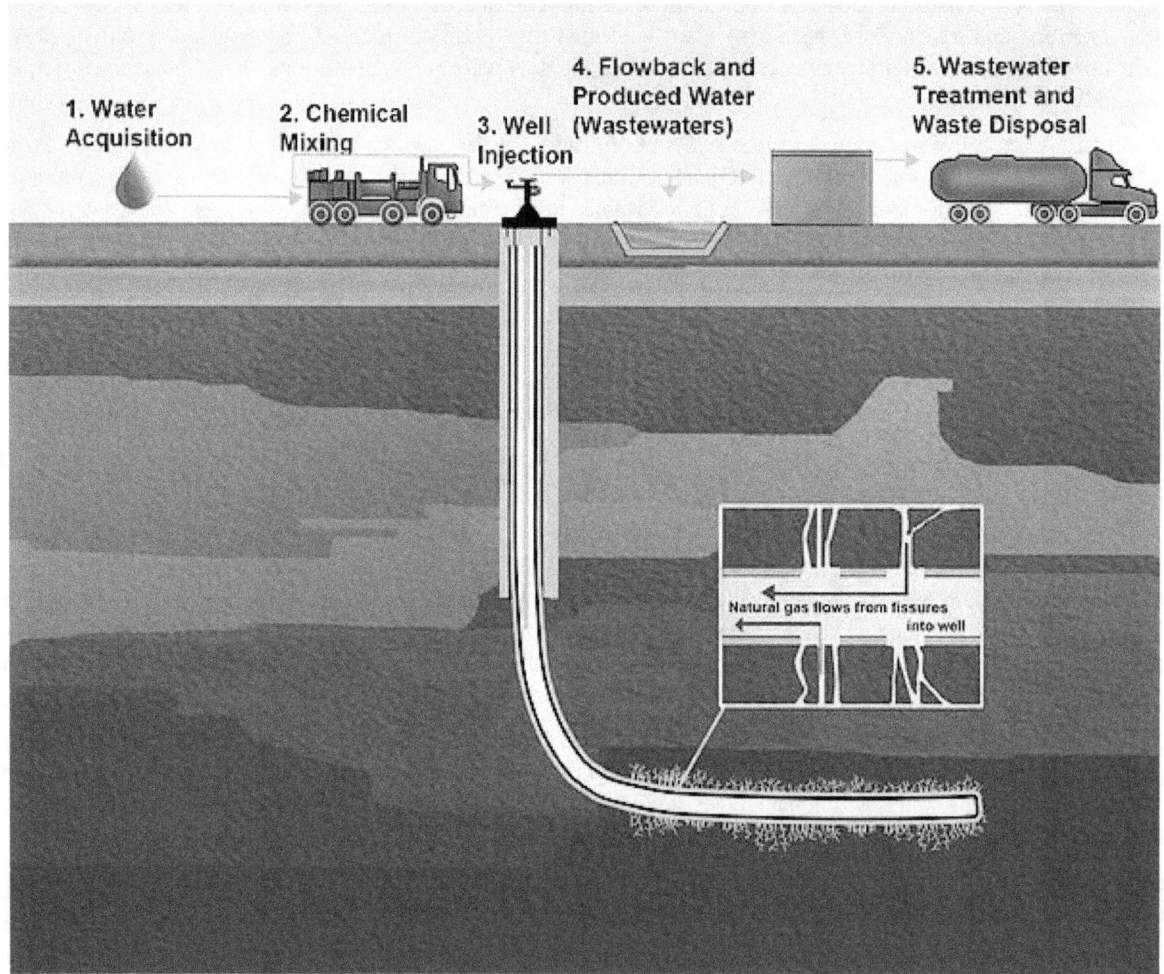

Illustration of hydraulic fracturing and related activities

The New York Times has reported that, since the 1980s, the EPA investigations into the oil and gas industry's environmental impact—including the ongoing one into fracking's potential impact on drinking water—and associated reports had been narrowed in scope and/or had negative findings removed due to industry and government pressure.[6][145]

A 2004 EPA study on hydraulic fracturing in coalbed methane wells concluded that the process was safe, and didn't warrant further study, because there was "no unequivocal evidence" of health risks to groundwater, and the fluids were neither necessarily hazardous nor able to travel far underground.[146] The EPA report did find uncertainties in knowledge of how fracturing fluid migrates through rocks, and recommended that diesel fuel not be used as a component of fracturing fluid in coalbed methane walls due to its potential as a source of benzene contamination; in response, well service companies agreed to stop using diesel fuel in coalbed methane wells.[147] One of the authors of the 2004 EPA report noted that it studied only hydraulic fracturing in coalbed methane wells.[146]

The New York Times cited Weston Wilson, the agency whistle-blower, that the results of the 2004 EPA study were influenced by industry and political pressure.[6] An early draft of the study discussed the possibility of dangerous levels of hydraulic fracturing fluid contamination and mentioned "possible evidence" of aquifer contamination. The final report concluded simply that hydraulic fracturing "poses little or no threat to drinking water".[6] The study's scope was narrowed so that it only focused on the injection of hydraulic fracturing fluids, ignoring other aspects of the process such as disposal of fluids and environmental concerns such as water quality, fish kills, and acid burns. The study was concluded before public complaints of contamination started emerging.[148]:780 The study's conclusion that the injection of hydraulic fracturing fluids into coalbed methane wells posed a minimal threat to underground drinking water sources[149] may have influenced the 2005 Congressional decision that hydraulic fracturing should continue to be regulated by the states and not under the federal Safe Drinking Water Act.

A 2011 study by Congressional Democrats and reporting by the New York Times that same year found that hydraulic fracturing had resulted in significant increases of radioactive material including radium and carcinogens including benzene in major rivers and watersheds.[150] At one site the amount of benzene discharged into the Allegheny River after treatment was 28 times accepted levels for drinking water.[150] The congressional representatives called for better regulation and more disclosure.[150]

In June 2015, the EPA released a report entitled "Assessment of the Potential Impacts of Hydraulic Fracturing for Oil and Gas on Drinking Water Resources" in which the EPA "did not find evidence that these mechanisms have led to widespread, systemic impacts on drinking water resources in the United States".[151] However, the EPA also noted that the mechanisms assessed in the report were not considered "widespread" and that evaluation of identified cases rests on limiting factors that include "insufficient pre- and post-fracturing data on the quality of drinking water resources; the paucity of long-term systematic studies; the presence of other sources of contamination precluding a definitive link between hydraulic fracturing activities and an impact; and the inaccessibility of some information on hydraulic fracturing activities and potential impacts."[151] The report suggested that two types of water withdrawals had potential for water resource contamination, namely ground water withdrawals and surface water withdrawals.[151] Perhaps more controversial is the recent Final Rule that was suspended on September 30, 2015 by US District Judge Scott Skavdahl with the Wyoming District Court.[152][153] Skavdahl entertained arguments that the regulative authority for hydraulic fracturing should rest with the EPA instead of the Bureau of Land Management.[152] Colorado, Utah (including the Ute Indian Tribe of the northern area of the state), Wyoming, North Dakota, the Independent Petroleum Association of America and the Western Energy Alliance included statements that the new rule would interfere in state regulations and cause redundancies that could take away resources from other programs.[152][153] Furthermore, Skavdahl considered the argument that the "final rules lack factual or scientific support" and that the opposition is supported by the recent publication of the June 2015 EPA report.[152]

1.6   Built Environment/Infrastructure

Hydraulic Fracturing's effects on built infrastructure are often underestimated. The fracking process requires heavy equipment and vast amount of water, chemicals, and other materials, thus transportation of that equipment, liquids, and materials, requires trucks with heavy tankers. This has caused infrastructure damage to local roads and bridges that were not designed and constructed to frequently withstand heavier loads.[154]

Each individual fracking well requires a vast amount of truck traffic. Studies estimated that on average, to fracture (build and drill) a single well, between 1,760 and 1,904 truck trips are needed to transport equipment, chemicals, water and other materials; removing fracking wastes and transporting the natural gas require additional truck trips.[155] The infrastructure deterioration caused by this heavy truck traffic has a huge economic impact/burden on local states. In July 2012, according to the Texas Department of Transportation, local fracking activities had cost an estimate of 2 billion dollars in damage to roads that connect drilling sites to storage sites.[156] In Pennsylvania, a study conducted in 2014 based on data on the distribution of fracking well activity and the roadway type in the state estimated that the road reconstruction costs caused by additional heavy truck traffic from Marcellus Shale natural gas development in 2011 were about $13,000–$23,000 per well for all state roadway types.[157]

Many similar studies are underway in different states to evaluate the potential infrastructure impact from fracking. However, existing evidence suggests that road and bridge deterioration from overloading infrastructure be taken into consideration when evaluating the environmental and economic cost of the fracking process.

1.7 See also

- Environmental issues in the United States

- Exemptions for hydraulic fracturing under United States federal law

- Pollution in the United States

1.8 References

[1] http://thinkprogress.org/climate/2014/10/22/3582904/methane-leaks-climate-benefit-fracking/

[2] "Chemicals Used in Hydraulic Fracturing" (PDF). Committee on Energy and Commerce U.S. House of Representatives. April 18, 2011.

[3] Brown, Valerie J. (February 2007). "Industry Issues: Putting the Heat on Gas". *Environmental Health Perspectives* (US National Institute of Environmental Health Sciences) **115** (2): A76. doi:10.1289/ehp.115-a76. PMC 1817691. PMID 17384744.

[4] Fischetti, Mark (August 20, 2013). "Groundwater Contamination May End the Gas-Fracking Boom". *Scientific American* **309** (3).

[5] Mall, Amy (19 December 2011). "Incidents where hydraulic fracturing is a suspected cause of drinking water contamination". *Switchboard: NRDC Staff Blog*. Natural Resources Defense Council. Retrieved 23 February 2012.

[6] Urbina, Ian (3 March 2011). "Pressure Limits Efforts to Police Drilling for Gas". *The New York Times*. Retrieved 23 February 2012. More than a quarter-century of efforts by some lawmakers and regulators to force the federal government to police the industry better have been thwarted, as E.P.A. studies have been repeatedly narrowed in scope and important findings have been removed

[7] http://www.mintpressnews.com/2500-square-mile-methane-plume-silently-hovering-western-us/200313/

[8] http://www.commondreams.org/news/2014/12/30/2500-square-mile-methane-plume-silently-hovering-over-western-us

[9] U.S. METHANE 'HOT SPOT' BIGGER THAN EXPECTED NASA, 9 Oct. 2014.

[10] "Ozone mitigation efforts continue in Sublette County, Wyoming". Wyoming's Online News Source. March 2011.

[11] Biello, David (30 March 2010). "Natural gas cracked out of shale deposits may mean the U.S. has a stable supply for a century—but at what cost to the environment and human health?". *Scientific American*. Retrieved 23 March 2012.

[12] Rodriguez, Ginna (April 2013). Air Emissions Characterization and Management For Natural Gas Hydraulic Fracturing Operations In the United States (PDF) (Report). University of Michigan School of Natural Resources and Environment. Retrieved 4 May 2014.

[13] "Oil and Natural Gas Air Pollution Standards". United States Environmental Protection Agency. Retrieved 2013-10-02.

[14] http://www.rt.com/usa/methane-emissions-fracking-underestimated-epa-024/

[15] https://www.washingtonpost.com/apps/g/page/national/unexpected-loose-gas-from-fracking/950/

[16] U.S. Environmental Protection Agency (2015). "Review of State and Industry Spill Data: Characterization of Hydraulic Fracturing-Related Spills". Retrieved 2015-10-10.

[17] U.S. Environmental Protection Agency (2015). "Assessment of the Potential Impacts of Hydraulic Fracturing for Oil and Gas on Drinking Water Resources".

[18] Andrews, Anthony; et al. (30 October 2009). Unconventional Gas Shales: Development, Technology, and Policy Issues (PDF) (Report). Congressional Research Service. pp. 7; 23. Retrieved 22 February 2012.

[19] Abdalla, Charles W.; Drohan, Joy R. (2010). Water Withdrawals for Development of Marcellus Shale Gas in Pennsylvania. Introduction to Pennsylvania's Water Resources (PDF) (Report). The Pennsylvania State University. Retrieved 16 September 2012. Hydrofracturing a horizontal Marcellus well may use 4 to 8 million gallons of water, typically within about 1 week. However, based on experiences in other major U.S. shale gas fields, some Marcellus wells may need to be hydrofractured several times over their productive life (typically five to twenty years or more)

[20] Ground Water Protection Council; ALL Consulting (April 2009). Modern Shale Gas Development in the United States: A Primer (PDF) (Report). DOE Office of Fossil Energy and National Energy Technology Laboratory. pp. 56–66. DE-FG26-04NT15455. Retrieved 24 February 2012.

[21] Arthur, J. Daniel; Uretsky, Mike; Wilson, Preston (May 5–6, 2010). *Water Resources and Use for Hydraulic Fracturing in the Marcellus Shale Region* (PDF). Meeting of the American Institute of Professional Geologists. Pittsburgh: ALL Consulting. p. 3. Retrieved 2012-05-09.

[22] Cothren, Jackson. Modeling the Effects of Non-Riparian Surface Water Diversions on Flow Conditions in the Little Red Watershed (PDF) (Report). U. S. Geological Survey, Arkansas Water Science Center Arkansas Water Resources Center, American Water Resources Association, Arkansas State Section Fayetteville Shale Symposium 2012. p. 12. Retrieved 16 September 2012. ...each well requires between 3 and 7 million gallons of water for hydraulic fracturing and the number of wells is expected to grow in the future

[23] Satterfield, J; Mantell, M; Kathol, D; Hiebert, F; Patterson, K; Lee, R (September 2008). *Managing Water Resources Challenges in Select Natural Gas Shale Plays.* GWPC Annual Meeting. ALL Consulting.

[24] "Unconventional well drilling permits". *Marcellus Center*. Marcellus Center, Pennsylvania State University. 2012. Retrieved 2012-09-16.

[25] "Horizontal drilling boosts Pennsylvania's natural gas production". EIA. 23 May 2012. Retrieved 2012-09-16.

[26] Ridlington, Elizabeth; John Rumpler (October 3, 2013). "Fracking by the numbers". *Environment America.*

[27] Lubber, Mindy (28 May 2013). "Escalating Water Strains In Fracking Regions". Forbes. Retrieved 20 October 2013.

[28] "A Texan tragedy: ample oil, no water" 11 Aug Guardian

[29] Berner, Daniel P; Grauman, Edward M; Hansen, Karen M; Kadas, Madeleine Boyer; LaValle, Laura L; Moore, Bryan J (May 1, 2013). "New Hydraulic Fracturing Water Recycling Rules Published in Texas Register". *The National Law Review* (Beveridge & Diamond PC). Retrieved 10 May 2013.

[30] Ridlington, Rumpler "Fracking by the numbers: key impact of dirty drilling at the state and national level", *Environment America*, October 2013

[31] Kris Fitz Patrick (November 17, 2011). "Ensuring Safe Drinking Water in the Age of Hydraulic Fracturing". The most fundamental recommendation is for states to rigorously test their ground water before and after hydraulic fracturing takes place. A major difficulty in proving or disproving contamination in previous cases has been the lack of a baseline sample for the water supply in question. The group also raises a federal policy issue, namely whether fracturing fluids should continue to be exempt from Safe Drinking Water Act regulations. This exemption was an informal one until 2005, when it was codified as part of the Energy Policy Act. A consequence of this exemption is that drilling companies are not required to disclose the chemicals that make up the fracturing fluids, making testing for these chemicals in ground water more difficult.

[32] Hass, Benjamin (14 August 2012). "Fracking Hazards Obscured in Failure to Disclose Wells". *Bloomberg News*. Retrieved 27 March 2013.

[33] Soraghan, Mike (13 December 2013). "White House official backs FracFocus as preferred disclosure method". *E&E News*. Retrieved 27 March 2013.

[34] "Colorado Sets The Bar on Hydraulic Fracturing Chemical Disclosure". Environmental Defense Fund. Retrieved 27 March 2013.

[35] Maykuth, Andrew (22 January 2012). "More states ordering disclosure of fracking chemicals.". *Philadelphia Inquirer*. Retrieved 27 March 2013.

[36] "Regulatory Statements on Hydraulic Fracturing Submitted by the States, June 2009" (PDF). Insterstate Oil and Gas Compact Commission. Retrieved 27 March 2013.

[37] "Pathways To Energy Independence: Hydraulic Fracturing And Other New Technologies". U.S. Senate. May 6, 2011.

[38] Ikeda, Robin (April 26, 2013). "Review of Federal Hydraulic Fracturing Research Activities. Testimony before the Subcommittees on Energy and Environment Committee on Science, Space and Technology U.S. House of Representatives". *CDC web site*. US Center for Disease Control and Prevention. Retrieved May 11, 2013.

[39] Urbina, Ian (3 August 2011). "A Tainted Water Well, and Concern There May be More". *The New York Times*. Retrieved 22 February 2012.

[40] "EPA's Study of Hydraulic Fracturing and Its Potential Impact on Drinking Water Resources". EPA. Retrieved 24 February 2010.

[41] Horwitt, Dusty (August 3, 2011). Cracks in the Facade: 25 Years Ago, EPA Linked "Fracking" to Water Contamination (PDF). *Environmental Working Group* (Report). Retrieved August 3, 2011.

[42] Brown, VJ (Feb 2007). "Industry Issues: Putting the Heat on Gas". *Environmental Health Perspectives* **115** (2): A76. doi:10.1289/ehp.115-a76. PMC 1817691. PMID 17384744. Retrieved 30 January 2015.

[43] "Timeline for cleanup from Clark gas well blowout accelerated". Casper Star Tribune. Associated Press. 27 Feb 2008. Retrieved 30 January 2015.

[44] Michael Rubinkam, Pa. regulators shut down Cabot drilling, April 15, 2010, pressconnects.com

[45] Lustgarten, Abrahm (November 20, 2009). "Pa. Residents Sue Gas Driller for Contamination, Health Concerns". *Pro Publica*. Retrieved February 4, 2014.

[46] Fetzer, Richard M. (January 19, 2012). Action Memorandum - Request for funding for a Removal Action at the Dimock Residential Groundwater Site (PDF) (Report). Retrieved May 27, 2012.

[47] Legere, Laura. "Gas company slapped with drilling ban and fine". The Times Tribune. Retrieved May 8, 2011.

[48] Mouawad, Jad; Krauss, Clifford (7 December 2009). "Dark Side of a Natural Gas Boom". *The New York Times*. Retrieved 3 March 2012.

[49] Christopher Bateman (21 June 2010). "A Colossal Fracking Mess". VanityFair.com. Retrieved 3 March 2012.

[50] Jim Snyder; Mark Drajem (10 January 2012). "Pennsylvania Fracking Foes Fault EPA Over Tainted Water Response". Bloomberg. Retrieved 19 January 2012.

[51] Gardner, Timothy (2012-05-11). "Water safe in town made famous by fracking-EPA". *Reuters*. Retrieved 2012-05-14.

[52] "Dimock, PA Water Testing Results Expected To Impact Fracking Debate". Associated Press. 5 March 2012. Retrieved 27 May 2012.

[53] DiGiulio, Dominic C.; Wilkin, Richard T.; Miller, Carlyle; Oberley, Gregory (December 2011). Investigation of Ground Water Contamination near Pavillion, Wyoming. Draft (PDF) (Report). EPA. Retrieved 23 March 2012.

[54] "EPA Releases Draft Findings of Pavillion, Wyoming Ground Water Investigation for Public Comment and Independent Scientific Review" (Press release). EPA. 8 December 2011. Retrieved 27 February 2012.

[55] Phillips, Susan (8 December 2011). "EPA Blames Fracking for Wyoming Groundwater Contamination". *StateImpact Pennsylvania* (NPR). Retrieved 6 February 2012.

[56] Peter R. Wright, Peter B. McMahon, David K. Mueller, and Melanie L. Clark (9 March 2012). Groundwater quality and quality control data for two monitoring wells near Pavillion, Wyoming, April and May 2012. (PDF) (Report). U.S. Geological Survey. Retrieved 29 September 2012.

[57] US EPA, Region 8, Wyoming to Lead Further Investigation of Water Quality Concerns Outside of Pavillion with Support of EPA, 20 June 2013.

[58] General Accounting Office, Energy-water nexus, p.15-17, 9 Jan. 2012.

[59] Kiparsky, Michael; Hein, Jayni Foley (April 2013). "Regulation of Hydraulic Fracturing in California: A Wastewater and Water Quality Perspective" (PDF). University of California Center for Law, Energy, and the Environment. Retrieved 2014-05-01.

[60] U.S. Environmental Protection Agency, Environmental Crimes Case Bulletin, Feb. 2013, p.10.

[61] US Environmental Protection Agency, Natural gas drilling in the Marcellus Shale: NPDES program FAQs, PDF, 16 March 2011.

[62] University of Pittsburgh, Shales Gas Roundtable, p.56, Aug, 2013.

[63] Lutz, Brian; Lewis, Aurana; Doyle, Martin (8 February 2013). "Generation, transport, and disposal of wastewater associated with Marcellus Shale gas development". *Environmental Health Perspectives* (Water Resources Research) **49** (2): 647–1197. doi:10.1002/wrcr.20096. Retrieved 2013-06-30.

[64] Logan, Jeffrey (2012). Natural Gas and the Transformation of the U.S. Energy Sector: Electricity (PDF) (Report). Joint Institute for Strategic Energy Analysis. Retrieved 27 March 2013.

[65] Alex Wayne (4 January 2012). "Health Effects of Fracking Need Study, Says CDC Scientist". *Businessweek*. Retrieved 29 February 2012.

[66] David Wethe (19 January 2012). "Like Fracking? You'll Love 'Super Fracking'". *Businessweek*. Retrieved 22 January 2012.

[67] Mark Drajem (11 January 2012). "Fracking Political Support Unshaken by Doctors' Call for Ban". Bloomberg. Retrieved 19 January 2012.

[68] "Environmental Defense Center: Fracking". Retrieved 22 April 2013.

[69] "ENERGY POLICY ACT OF 2005" (PDF). *Authentic Government Information GPO*. Retrieved 23 April 2013. (i) the underground injection of natural gas for purposes of storage; and (ii) the underground injection of fluids or propping agents (other than diesel fuels) pursuant to hydraulic fracturing operations related to oil, gas, or geothermal production activities

[70] "With Natural Gas Drilling Boom, Pennsylvania Faces an Onslaught of Wastewater". Propublica. 3 October 2009. Retrieved 7 August 2013.

[71] Urbina, Ian (26 February 2011). "Regulation Lax as Gas Wells' Tainted Water Hits Rivers". *The New York Times*. Retrieved 22 February 2012.

[72] Caruso, David B. (2011-01-03). "44,000 Barrels of Tainted Water Dumped Into Neshaminy Creek. We're the only state allowing tainted water into our rivers". NBC Philadelphia. Associated Press. Retrieved 2012-04-28. ...the more than 300,000 residents of the 17 municipalities that get water from the creek or use it for recreation were never informed that numerous public pronouncements that the watershed was free of gas waste had been wrong.

[73] Kevin Begos (5 January 2014). "4 states confirm water pollution from drilling. Associated Press review of complaints casts doubt on industry view that it rarely happens.". *USA Today*. Associated Press. Retrieved 6 January 2014.

[74] Bruce Gellerman and Ann Murray (10 August 2012). "Disposal of Fracking Wastewater Polluting PA Rivers". *PRI's Environmental News Magazine* (Public Radio International). Retrieved 14 January 2013.

[75] Sun, M.; Lowry, G.V.; Gregory, K.B. (2013). "Selective oxidation of bromide in wastewater brines from hydraulic fracturing". *Water research* (Water Res.) **47** (11): 3723–3731. doi:10.1016/j.watres.2013.04.041. PMID 23726709.

[76] Paul Handke, Trihalomethane speciation and the relationship to elevated total dissolved solid concentrations, Pennsylvania Department of Environmental Protection.

[77] White, Jeremy; Park, Haeyoun; Urbina, Ian; Palmer, Griff (26 February 2011). "Toxic Contamination From Natural Gas Wells". *The New York Times*.

[78] "Drilling Down: Documents: Natural Gas's Toxic Waste". *The New York Times*. 26 February 2011. Retrieved 23 February 2012.

[79] Carus, Felicity (2 October 2013). "Dangerous levels of radioactivity found at fracking waste site in Pennsylvania. Co-author of study says UK must impose better environmental regulation than US if it pursues shale gas extraction". The Guardian. Retrieved 10 October 2013.

[80] Warner, Nathaniel R.; Christie, Cidney A.; Jackson, Robert B.; Vengosh, Avner (2 October 2013). "Impacts of Shale Gas Wastewater Disposal on Water Quality in Western Pennsylvania". *Environ. Sci. Technol.* (ACS Publications) **47** (20): 11849–57. doi:10.1021/es402165b. PMID 24087919.

[81] Jacobs, Harrison (9 October 2013). "Duke Study: Fracking Is Leaving Radioactive Pollution In Pennsylvania Rivers". *Business Insider* (Business Insider). Retrieved 10 October 2013.

[82] Griswold, Eliza (17 November 2011). "The Fracturing of Pennsylvania". *The New York Times Magazine*. Retrieved 21 November 2011.

[83] "State Official: Pa. Water Meets Safe Drinking Standards". CBS Pittsburgh. January 4, 2011.

[84] "Pennsylvania DEP Secretary Defends States' Ability to Regulate Hydraulic Fracturing". PR Newswire. November 17, 2011.

[85] Don Hopey (February 24, 2011). "Corbett repeals policy on gas drilling in parks". *Pittsburgh Post-Gazette*. Retrieved April 19, 2011.

[86] Bill McKibben (8 March 2012). "Why Not Frack?". *The New York Review of Books* **59** (4). Retrieved 21 February 2012.

[87] "Annual Drinking Water Quality Report, 2010" (PDF). Philadelphia Water Department. Spring 2011. Retrieved 7 February 2012.

[88] McGraw, Seamus (27 March 2011). "Is Fracking Safe? The Top 10 Myths About Natural Gas Drilling". *Popular Mechanics*. Retrieved 27 March 2013.

[89] Urbina, Ian (7 April 2011). "Pennsylvania Calls for More Water Tests". *The New York Times*. Retrieved 23 February 2012.

[90] "Natural Gas Drilling, the Spotlight". *The New York Times*. 5 March 2011. Retrieved 24 February 2012.

[91] Charles Petit (2 March 2011). "Part II of the fracking water problems in PA and other Marcellus Shale country". *Knight Science Journalism Tracker*. MIT. Retrieved 24 February 2012.

[92] "Analysis of Marcellus flowback finds high levels of ancient brines" (Press release). Pennsylvania State University. 17 December 2012. Retrieved 31 January 2013.

[93] Don Hopey, Study finds lower bromide levels in Mon, but not in Allegheny, Pittsburgh Post-Gazette, 13 Nov. 2012.

[94] Urbina, Ian (1 March 2011). "Drilling Down: Wastewater Recycling No Cure-All in Gas Process". *The New York Times*. Retrieved 22 February 2012.

[95] Don Hopey (5 March 2011). "Radiation-fracking link sparks swift reactions". *Pittsburgh Post-Gazette*. Retrieved 23 February 2012.

[96] Shocker: New York Times radioactive water report is false March 8, 2011 ι Abby Wisse Schachter. Report is from a Rupert Murdoch tabloid, The New York Post

[97] Urbina, Ian (7 March 2011). "E.P.A. Steps Up Scrutiny of Pollution in Pennsylvania Rivers". *The New York Times*. Retrieved 14 May 2013.

[98] Reis, John C. (1976). *Environmental Control in Petroleum Engineering*. Gulf Professional Publishers.

[99] K. Fisher and others, "A comprehensive study of the analysis and economic benefits of radioactive tracer engineered stimulation procedures," *Society of Petroleum Engineers*, Paper 30794-MS, October 1995.

[100] Jack E. Whitten, Steven R. Courtemanche, Andrea R. Jones, Richard E. Penrod, and David B. Fogl (Division of Industrial and Medical Nuclear Safety, Office of Nuclear Material Safety and Safeguards (June 2000). "Consolidated Guidance About Materials Licenses: Program-Specific Guidance About Well Logging, Tracer, and Field Flood Study Licenses (NUREG-1556, Volume 14)". US Nuclear Regulatory Commission. Retrieved 19 April 2012. labeled Frac Sand...Sc-46, Br-82, Ag-110m, Sb-124, Ir-192

[101] Radiation Protection and the Management of Radioactive Waste in the Oil and Gas Industry (PDF) (Report). International Atomic Energy Agency. 2003. pp. 39–40. Retrieved 20 May 2012.

[102] Timothy A. Bartrand and Jeffrey S. Rosen (October 2013). Potential Impacts and Significance of Elevated 131 I on Drinking Water Sources [Project #4486] ORDER NUMBER: 4486 (PDF) (Report). Water Research Foundation. Retrieved 11 November 2013.

[103] Bennet, Les, *et.al.*. "The Source for Hydraulic Fracture Characterization" (PDF). *Oilfield Review* (Schlumberger) (Winter 2005/2006): 42–57. Retrieved 2012-09-30.

[104] Ellsworth, W. L.; Hickman, S.H.; McGarr, A.; Michael, A. J.; Rubinstein, J. L. (18 April 2012). *Are seismicity rate changes in the midcontinent natural or manmade?*. Seismological Society of America 2012 meeting. San Diego, California: Seismological Society of America. Retrieved 2014-02-23.

[105] US Geological Survey, Man-made earthquakes, accessed 22 Sept. 2013.

[106] Zoback, Mark; Kitasei, Saya; Copithorne, Brad (July 2010). Addressing the Environmental Risks from Shale Gas Development (PDF) (Report). Worldwatch Institute. p. 9. Retrieved 2012-05-24.

[107] US Geological Survey, Hydraulic fracturing FAQs, accessed 21 April 2015.

[108] Soraghan, Mike (13 December 2013). "Disconnects in public discourse around 'fracking' cloud earthquake issue.". *E&E News*. Retrieved 27 March 2013.

[109] Induced Seismicity Potential in Energy Technologies (Report). National Academies Press. 2012. Retrieved 27 March 2013. The process of hydraulic fracturing a well as presently implemented for shale gas recovery does not pose a high risk for inducing felt seismic events.

[110] Kim, Won-Young 'Induced seismicity associated with fluid injection into a deep well in Youngstown, Ohio', Journal of Geophysical Research-Solid Earth

[111] van der Elst1, Nicholas J.; Savage, Heather M.; Keranen, Katie M; Abers, Geoffrey A. (12 July 2013). "Enhanced Remote Earthquake Triggering at Fluid-Injection Sites in the Midwestern United States". *Science* (ACS Publications) **341** (6142): 164–167. doi:10.1126/science.1238948. PMID 23846900.

[112] Frohlich, Cliff (2012). "Two-year survey comparing earthquake activity and injection-well locations in the Barnett Shale, Texas.". *Proceedings of the National Academy of Sciences of the United States of America* **109**: 13934–13938. doi:10.1073/pnas.1207728109.

[113] Cliff Frohlich, Induced or triggered Earthquakes in Texas, Final Technical Report, Award no. G12AP20001, US Geological Survey, External Report, n.d.

[114] , Environmental Protection Agency

[115] USGS, How large are the earthquakes induced by fluid injection?

[116] "How is hydraulic fracturing related to earthquakes and tremors?". USGS. Retrieved 4 November 2012.

[117] Rachel Maddow, Terrence Henry (7 August 2012). *Rachel Maddow Show: Fracking waste messes with Texas* (video). MSNBC. Event occurs at 9:24 - 10:35.

[118] Soraghan, Mike (29 March 2012). "'Remarkable' spate of man-made quakes linked to drilling, USGS team says". *EnergyWire* (E&E). Retrieved 2012-11-09.

[119] Henry, Terrence (6 August 2012). "How Fracking Disposal Wells Are Causing Earthquakes in Dallas-Fort Worth". *State Impact Texas*. NPR. Retrieved 9 November 2012.

[120] Katie M. Keranen, "Potentially induced earthquakes in Oklahoma, USA," *Geology*, 26 March 2013.

[121] Justin L. Rubenstein, "Myths and facts on wastewater injection, hydraulic fracturing, and induced seismicity,", *Seismological Research Letters*, 10 June 2015.

[122] Bill Leith, Induced seismicity, US Geological Survey, June 2012.

[123] "Ohio Quakes Probably Triggered by Waste Disposal Well, Say Seismologists" (Press release). Lamont–Doherty Earth Observatory. 6 January 2012. Retrieved 22 February 2012.

[124] Ohio Department of Natural Resources, Executive Summary, Preliminary Report on the Northstar 1 Class II Injection Well and the Seismic Events in the Youngstown, Ohio, Area, PDF, March 2012.

[125] NPR - State Impact Texas, How oil and gas disposal wells can cause earthquakes.

[126] University of Texas, Study finds correlation between injection wells and small earthquakes, 6 Aug. 2011.

[127] Dunn, Sharon (June 5, 2014). "CU research team studying earthquake activity near Greeley". *Greeley Tribune*. Retrieved 10 June 2014.

[128] Tomasic, John (June 2, 2014). "Greeley quake adds ammunition to Colorado fracking war". *The Colorado Independent*. Retrieved 10 June 2014.

[129] Mary Kang and others, "Direct measurements of methane emissions from abandoned oil and gas wells in Pennsylvania", Proceedings of the National Academy of Science, 23 Dec. 2014, v.11, n.51, p18173-18177.

[130] http://www.scientificamerican.com/article/abandoned-wells-leak-powerful-greenhouse-gas/

[131] http://www.climatecentral.org/news/abandoned-oil-wells-methane-emissions-17575

[132] Finkel ML, Hays J; Hays (October 2013). "The implications of unconventional drilling for natural gas: a global public health concern". *Public Health* (Review) **127** (10): 889–93. doi:10.1016/j.puhe.2013.07.005. PMID 24119661.

[133] Adgate, John L.; Goldstein, Bernard D.; McKenzie, Lisa M. (2014-08-05). "Potential Public Health Hazards, Exposures and Health Effects from Unconventional Natural Gas Development". *Environmental Science & Technology* **48** (15): 8307–8320. doi:10.1021/es404621d. ISSN 0013-936X.

[134] McKenzie, Lisa M.; Guo, Ruixin; Witter, Roxana Zulauf; Savitz, David A.; Newman, Lee S.; Adgate, John L. "Birth Outcomes and Maternal Residential Proximity to Natural Gas Development in Rural Colorado". *Environmental Health Perspectives*. doi:10.1289/ehp.1306722. PMC 3984231. PMID 24474681.

[135] McKenzie, Lisa M.; Witter, Roxana Z.; Newman, Lee S.; Adgate, John L. (2012-05-01). "Human health risk assessment of air emissions from development of unconventional natural gas resources". *Science of The Total Environment* **424**: 79–87. doi:10.1016/j.scitotenv.2012.02.018.

[136] (PDF) http://www2.epa.gov/sites/production/files/2015-06/documents/hf_es_erd_jun2015.pdf. Missing or empty |title= (help)

[137] Stacy, Shaina L.; Brink, LuAnn L.; Larkin, Jacob C.; Sadovsky, Yoel; Goldstein, Bernard D.; Pitt, Bruce R.; Talbott, Evelyn O. (2015-06-03). "Perinatal Outcomes and Unconventional Natural Gas Operations in Southwest Pennsylvania". *PLoS ONE* **10** (6): e0126425. doi:10.1371/journal.pone.0126425. PMC 4454655. PMID 26039051.

[138] Eaton TT. Science-based decision-making on complex issues: Marcellus shale gas hydrofracking and New York City water supply. Sci Total Environ. 2013 Sep 1;461-462:158-69. doi: 10.1016/j.scitotenv.2013.04.093. Epub 2013 May 28. PMID 23722091

[139] Mall, Amy (16 May 2012). "Concerns about the health risks of fracking continue to grow". *Switchboard: NRDC Staff Blog*. Natural Resources Defense Council. Retrieved 2012-05-19.

[140] Hopkinson, Jenny; DiCosmo, Bridget (15 May 2012). "Academies' NRC Seeks Broad Review Of Currently Ignored Fracking Risks". *InsideEPA* (Inside Washington Publishers). (subscription required). Retrieved 2012-05-19.

[141] Health Consultation, Garfield County, Colorado, US Agency for Toxic Substances and Disease Registry, 13 March 2015, p.10 and Table 2.

[142] Abrahm Lustgarten and Nicholas Kusnetz (2011-09-16). "Science Lags as Health Problems Emerge Near Gas Fields". Propublica. Retrieved 2013-05-06.

[143] "Worker Exposure to Silica during Hydraulic Fracturing". OSHA. Retrieved 15 January 2013.

[144] Esswein, Eric; Kiefer, Max; Snawder, John; Breitenstein, Michael (23 May 2012). "Worker Exposure to Crystalline Silica During Hydraulic Fracturing". *NIOSH Science Blog*. United States Center for Disease Control. Retrieved 2012-09-08.

[145] "The Debate Over the Hydrofracking Study's Scope". *The New York Times*. 3 March 2011. Retrieved 1 May 2012. While environmentalists have aggressively lobbied the agency to broaden the scope of the study, industry has lobbied the agency to narrow this focus

[146] "Does Natural-Gas Drilling Endanger Water Supplies?". *BusinessWeek*. November 11, 2008.

[147] "Evaluation of Impacts to Underground Sources of Drinking Water by Hydraulic Fracturing of Coalbed Methane Reservoirs; National Study Final Report" (PDF). Retrieved July 13, 2011.

[148] Dammel, Joseph A. (2011). "Notes From Underground: Hydraulic Fracturing in the Marcellus Shale" (PDF). *Minnesota Journal of Law, Science and Technology* (University of Minnesota Law School) **12** (2): 773–810. Retrieved 24 February 2012.

[149] Evaluation of Impacts to Underground Sources of Drinking Water by Hydraulic Fracturing of Coalbed Methane Reservoirs; National Study Final Report (PDF) (Report). EPA. June 2004. Retrieved 23 February 2011.

[150] Urbina, Ian (April 16, 2011). "Chemicals Were Injected Into Wells, Report Says". *New York Times.* Retrieved May 2, 2011.

[151] EPA. "Assessment of the Potential Impacts of Hydraulic Fracturing for Oil and Gas on Drinking Water Resources" (PDF). *United States Environmental Protection Agency.* EPA. Retrieved 28 October 2015.

[152] Haun, Marjorie (8 October 2015). "Federal judge gives fracking a break from BLM regulations". Watchdog Arena. Retrieved 28 October 2015.

[153] Associated Press (24 June 2015). "At The Last Minute, Judge Delays Federal Fracking Regulations". Colorado Public Radio. Retrieved 28 October 2015.

[154] Mehany, M.S.H.M.; Guggemos, A. (2015). "A Literature Survey of the Fracking Economic and Environmental Implications in the United States" (PDF). *Procedia Engineering* (118): 169–176. Retrieved 10/25/15. Check date values in: |access-date= (help)

[155] Podulka, S.G.; Podulka, W.J. (6/9/2010). "Comments on the Science Advisory Board's 5/19/2010 Draft Committee Report on the EPA's Research Scoping Document Related to Hydraulic Fracturing ("Report")" (PDF). *EPA Science Advisor Board.* Retrieved 26 October 2015. Check date values in: |date= (help)

[156] Shlachter, B. "Drilling trucks have caused an estimated $2 billion in damage to Texas roads.". *Star-Telegram.* Retrieved 26 October 2015.

[157] Abramzon, S; Samaras, C; Curtright, A; Litovitz, A; Burger, N (2014). "Estimating the Consumptive Use Costs of Shale Natural Gas Extraction on Pennsylvania Roadways". *Journal of Infrastructure Systems* **20** (3). doi:10.1061/(ASCE)IS.1943-555X.0000203.

1.9   Further reading

- Joseph D. Ayotte; et al. (August 2011). "Trace Elements and Radon in Groundwater Across the United States, 1992-2003". U.S. Geological Survey. Retrieved 25 May 2012.

- Bamberger, Michelle; Oswald, Robert E. (2012). "Impacts of gas drilling on human and animal health" (PDF). *New Solutions: A Journal of Environmental and Occupational Health Policy* **22** (1): 51–77. doi:10.2190/NS.22.1.e. Retrieved 2012-12-21.

- Colorado Oil & Gas Conservation Commission. "Gasland Correction Document" (PDF). Retrieved 7 August 2013.

- Colborn, Theo; Kwiatkowski, Carol; Schultz, Kim; Bachran, Mary (2011). "Natural Gas Operations from a Public Health Perspective" (PDF). *Human and Ecological Risk Assessment: an International Journal* (Taylor & Francis) **17** (5): 1039–1056. doi:10.1080/10807039.2011.605662.

- DiCosmo, Bridget (15 May 2012). "SAB Pushes To Advise EPA To Conduct Toxicity Tests In Fracking Study". *InsideEPA.* Inside Washington Publishers. (subscription required). Retrieved 2012-05-19.

- Mark Drajem (27 September 2012). "Diesel in Water Near Fracking Confirms EPA Tests Wyoming Disputes". Bloomberg News. Retrieved 28 September 2012.

- Energy Institute (February 2012). Fact-Based Regulation for Environmental Protection in Shale Gas Development (PDF) (Report). University of Texas at Austin. Retrieved 29 February 2012.

- Fontenot, Brian E.; Hunt, Laura R.; Hildenbrand, Zacariah L.; Carlton Jr., Doug D.; Oka, Hyppolite; Walton, Jayme L. (2013). "An Evaluation of Water Quality in Private Drinking Water Wells Near Natural Gas Extraction Sites in the Barnett Shale Formation". *Environ. Sci. Technol.* **47** (17): 10032–10040. doi:10.1021/es4011724. PMID 23885945.

- Grant, Alison (4 April 2013). "FracTracker monitors shale development in Ohio". The Plain Dealer. Retrieved 28 July 2013.

- Mead Gruver (12 December 2011). "New Data, but Not Much New in Wyo. Fracking Study". EPA. Retrieved 6 May 2013.

- Christopher Helman (8 December 2011). "What If Fracking Did Pollute Wyoming Water?". Forbes.com. Retrieved 6 February 2012.

- Jackson, Robert B.; Vengosh, Avner; Darrah, Thomas H.; Warner, Nathaniel R.; Down, Adrian; Poreda, Robert J.; Osborn, Stephen G.; Zhao, Kaiguang; Karr, Jonathan D. (24 June 2013). "Increased Stray Gas Abundance in a Subset of Drinking Water Wells Near Marcellus Shale Gas Extraction" (PDF). *Proceedings of the National Academy of Sciences* **110** (28): 11250–11255. doi:10.1073/pnas.1221635110. Retrieved 2014-03-30.

- McKenzie, Lisa; Witter, Roxana; Newman, Lee; Adgate, John (2012). "Human health risk assessment of air emissions from development of unconventional natural gas resources". *Science of the Total Environment* **424**: 79–87. doi:10.1016/j.scitotenv.2012.02.018. PMID 22444058.

- Molofsky, L.J.; Connor, J.A.; Shahla, K.F.; Wylie, A.S.; Wagner, T. (December 5, 2011). "Methane in Pennsylvania Water Wells Unrelated to Marcellus Shale Fracturing". *Oil and Gas Journal* (Pennwell Corporation) **109** (49): 54–67.

- Moniz, Ernest J.; et al. (June 2011). The Future of Natural Gas: An Interdisciplinary MIT Study (PDF) (Report). Massachusetts Institute of Technology. Retrieved 1 June 2012.

- Munro, Margaret (17 February 2012). "Fracking does not contaminate groundwater: study released in Vancouver". *Vancouver Sun*. Retrieved 3 March 2012.

- Osborn, Stephen G.; Vengosh, Avner; Warner, Nathaniel R.; Jackson, Robert B. (2011-05-17). "Methane contamination of drinking water accompanying gas-well drilling and hydraulic fracturing" (PDF). *Proceedings of the National Academy of Sciences of the United States of America* **108** (20): 8172–8176. doi:10.1073/pnas.1100682108. Retrieved 2011-10-14.

- phillynowstaff (9 December 2011). "EPA Releases Report on Water Contamination By Fracking, As GA Pushes Fee Bills". *PhillyNow blog*. Philadelphia Weekly. Archived from the original on 16 July 2012. Retrieved 6 February 2012.

- PEHSU (August 2011). PEHSU Information Concerning Effects on Children of Natural Gas Extraction and Hydraulic Fracturing (Report). Propublica. Retrieved 2013-05-06.

- Schmidt, Charles W. (August 2011). "Blind Rush? Shale Gas Boom Proceeds Amid Human Health Questions". *Environmental Health Perspectives* **119** (119(1)): A348–53. doi:10.1289/ehp.119-a348. PMC 3237379. PMID 21807583.

- Vaughan, Vicki (16 February 2012). "Fracturing 'has no direct' link to water pollution, UT study finds". Retrieved 3 March 2012.

- Helen Westerman (11 January 2012). "Gas drilling research highlights risk to animals, but more thorough work needed". *The Conversation*. Retrieved 25 May 2012.

1.10 External links

- FracTracker.org Maps, data, and articles from news, government, industry, and academic sources.

- "FAQ: Hydraulic Fracturing, SDWA, Fluids, and DeGette/Casey" (PDF). Energy In Depth. Retrieved 27 March 2013.

- "Groundwater Investigation: Pavillion, WY". EPA. Retrieved 6 February 2012.

- *Natural gas wells leakier than believed*: Measurements at Colorado site show methane release higher than previous estimates 24 March 2012, differences between NOAA and United States Environmental Protection Agency estimates

Chapter 2

Hydraulic fracturing

"Fracking" redirects here. For other uses, see Frack (disambiguation) and Frac (disambiguation).
This article is about hydraulic fracturing on a global scale. For information specific to the United States, see Hydraulic fracturing in the United States and Environmental impact of hydraulic fracturing in the United States.

Hydraulic fracturing (also **hydrofracturing**, **hydrofracking**, **fracking** or **fraccing**) is a well-stimulation technique in which rock is fractured by a pressurized liquid. The process involves the high-pressure injection of 'fracking fluid' (primarily water, containing sand or other proppants suspended with the aid of thickening agents) into a wellbore to create cracks in the deep-rock formations through which natural gas, petroleum, and brine will flow more freely. When the hydraulic pressure is removed from the well, small grains of hydraulic fracturing proppants (either sand or aluminium oxide) hold the fractures open.[1]

Hydraulic fracturing began as an experiment in 1947, and the first commercially successful application followed in 1950. As of 2012, 2.5 million "frac jobs" had been performed worldwide on oil and gas wells; over one million of those within the U.S.[2][3] Such treatment is generally necessary to achieve adequate flow rates in shale gas, tight gas, tight oil, and coal seam gas wells.[4] Some hydraulic fractures can form naturally in certain veins or dikes.[5]

Hydraulic fracturing is highly controversial in many countries. Its proponents advocate the economic benefits of more extensively accessible hydrocarbons.[6][7] However, opponents argue that these are out-weighed by the potential environmental impacts, which include risks of ground and surface water contamination, air and noise pollution, and potentially triggering earthquakes, along with the consequential hazards to public health and the environment.[8][9]

Increases in seismic activity following hydraulic fracturing along dormant or previously unknown faults are sometimes caused by the deep-injection disposal of hydraulic fracturing flowback (a byproduct of hydraulically fractured wells),[10] and produced formation brine (a byproduct of both fractured and nonfractured oil and gas wells).[11] For these reasons, hydraulic fracturing is under international scrutiny, restricted in some countries, and banned altogether in others.[12][13][14] Some countries have banned the practice or put moratoria in place, while others have adopted an approach involving tight regulation. The European Union is drafting regulations that would permit controlled application of hydraulic fracturing.[15]

2.1 Geology

Main article: Fracture (geology)

2.1.1 Mechanics

Fracturing rocks at great depth frequently becomes suppressed by pressure due to the weight of the overlying rock strata and the cementation of the formation. This suppression process is particularly significant in "tensile" (Mode 1) fractures

Halliburton fracturing operation in the Bakken Formation, North Dakota, United States

which require the walls of the fracture to move against this pressure. Fracturing occurs when effective stress is overcome by the pressure of fluids within the rock. The minimum principal stress becomes tensile and exceeds the tensile strength of the material.[16][17] Fractures formed in this way are generally oriented in a plane perpendicular to the minimum principal stress, and for this reason, hydraulic fractures in well bores can be used to determine the orientation of stresses.[18] In natural examples, such as dikes or vein-filled fractures, the orientations can be used to infer past states of stress.[19]

2.1.2   Veins

Most mineral vein systems are a result of repeated natural fracturing during periods of relatively high pore fluid pressure. The impact of high pore fluid pressure on the formation process of mineral vein systems is particularly evident in "crack-seal" veins, where the vein material is part of a series of discrete fracturing events, and extra vein material is deposited on each occasion.[20] One example of long-term repeated natural fracturing is in the effects of seismic activity. Stress levels rise and fall episodically, and earthquakes can cause large volumes of connate water to be expelled from fluid-filled fractures. This process is referred to as "seismic pumping".[21]

2.1.3   Dikes

Minor intrusions in the upper part of the crust, such as dikes, propagate in the form of fluid-filled cracks. In such cases, the fluid is magma. In sedimentary rocks with a significant water content, fluid at fracture tip will be steam.[22]

A fracturing operation in progress

2.2 History

2.2.1 Precursors

Fracturing as a method to stimulate shallow, hard rock oil wells dates back to the 1860s. Dynamite or nitroglycerin detonations were used to increase oil and natural gas production from petroleum bearing formations. On April 25, 1865, Civil War veteran Col. Edward A. L. Roberts received a patent for an "exploding torpedo".[23] It was employed in Pennsylvania, New York, Kentucky, and West Virginia using liquid and also, later, solidified nitroglycerin. Later still the same method was applied to water and gas wells. Stimulation of wells with acid, instead of explosive fluids, was introduced in the 1930s. Due to acid etching, fractures would not close completely resulting in further productivity increase.[24]

2.2.2 Oil and gas wells

The relationship between well performance and treatment pressures was studied by Floyd Farris of Stanolind Oil and Gas Corporation. This study was the basis of the first hydraulic fracturing experiment, conducted in 1947 at the Hugoton gas field in Grant County of southwestern Kansas by Stanolind.[4][25] For the well treatment, 1,000 US gallons (3,800 l; 830 imp gal) of gelled gasoline (essentially napalm) and sand from the Arkansas River was injected into the gas-producing limestone formation at 2,400 feet (730 m). The experiment was not very successful as deliverability of the well did not change appreciably. The process was further described by J.B. Clark of Stanolind in his paper published in 1948. A patent on this process was issued in 1949 and exclusive license was granted to the Halliburton Oil Well Cementing Company. On March 17, 1949, Halliburton performed the first two commercial hydraulic fracturing treatments in Stephens County,

Oklahoma, and Archer County, Texas.[25] Since then, hydraulic fracturing has been used to stimulate approximately one million oil and gas wells[26] in various geologic regimes with good success.

In contrast with large-scale hydraulic fracturing used in low-permeability formations, small hydraulic fracturing treatments are commonly used in high-permeability formations to remedy "skin damage", a low-permeability zone that sometimes forms at the rock-borehole interface. In such cases the fracturing may extend only a few feet from the borehole.[27]

In the Soviet Union, the first hydraulic proppant fracturing was carried out in 1952. Other countries in Europe and Northern Africa subsequently employed hydraulic fracturing techniques including Norway, Poland, Czechoslovakia, Yugoslavia, Hungary, Austria, France, Italy, Bulgaria, Romania, Turkey, Tunisia, and Algeria.[28]

2.2.3 Massive fracturing

Massive hydraulic fracturing (also known as high-volume hydraulic fracturing) is a technique first applied by Pan American Petroleum in Stephens County, Oklahoma, USA in 1968. The definition of massive hydraulic fracturing varies somewhat, but is generally reference to treatments injecting greater than about 150 short tons, or approximately 300,000 pounds (136 metric tonnes), of proppant.[29]

American geologists became increasingly aware that there were huge volumes of gas-saturated sandstones with permeability too low (generally less than 0.1 millidarcy) to recover the gas economically.[29] Starting in 1973, massive hydraulic fracturing was used in thousands of gas wells in the San Juan Basin, Denver Basin,[30] the Piceance Basin,[31] and the Green River Basin, and in other hard rock formations of the western US. Other tight sandstone wells in the US made economically viable by massive hydraulic fracturing were in the Clinton-Medina Sandstone, and Cotton Valley Sandstone.[29]

Massive hydraulic fracturing quickly spread in the late 1970s to western Canada, Rotliegend and Carboniferous gas-bearing sandstones in Germany, Netherlands (onshore and offshore gas fields), and the United Kingdom in the North Sea.[28]

Horizontal oil or gas wells were unusual until the late 1980s. Then, operators in Texas began completing thousands of oil wells by drilling horizontally in the Austin Chalk, and giving massive *slickwater* hydraulic fracturing treatments to the wellbores. Horizontal wells proved much more effective than vertical wells in producing oil from tight chalk;[32] sedimentary beds are usually nearly horizontal, so horizontal wells have much larger contact areas with the target formation.[33]

2.2.4 Shales

Hydraulic fracturing of shales goes back at least to 1965, when some operators in the Big Sandy gas field of eastern Kentucky and southern West Virginia started hydraulically fracturing the Ohio Shale and Cleveland Shale, using relatively small fracs. The frac jobs generally increased production, especially from lower-yielding wells.[34]

In 1976, the United States government started the Eastern Gas Shales Project, which included numerous public-private hydraulic fracturing demonstration projects.[35] During the same period, the Gas Research Institute, a gas industry research consortium, received approval for research and funding from the Federal Energy Regulatory Commission.[36]

In 1997, taking the slickwater fracturing technique used in East Texas by Union Pacific Resources (now part of Anadarko Petroleum Corporation), Mitchell Energy (now part of Devon Energy), applied the technique in the Barnett Shale of north Texas.[33] This made gas extraction widely economical in the Barnett Shale, and was later applied to other shales.[37][38][39] George P. Mitchell has been called the "father of fracking" because of his role in applying it in shales.[40] The first horizontal well in the Barnett Shale was drilled in 1991, but was not widely done in the Barnett until it was demonstrated that gas could be economically extracted from vertical wells in the Barnett.[33]

As of 2013, massive hydraulic fracturing is being applied on a commercial scale to shales in the United States, Canada, and China. Several additional countries are planning to use hydraulic fracturing.[41][42][43]

Well head where fluids are injected into the ground

Well head after all the hydraulic fracturing equipment has been taken off location

2.3   Process

According to the United States Environmental Protection Agency (EPA) hydraulic fracturing is a process to stimulate a natural gas, oil, or geothermal energy well to maximize extraction. EPA defines the broader process as including the acquisition of source water, well construction, well stimulation, and waste disposal.[44]

2.3.1   Method

A hydraulic fracture is formed by pumping fracturing fluid into a wellbore at a rate sufficient to increase pressure at the target depth (determined by the location of the well casing perforations), to exceed that of the fracture *gradient* (pressure gradient) of the rock.[45] The fracture gradient is defined as pressure increase per unit of depth relative to density, and is usually measured in pounds per square inch, per square foot, or bars. The rock cracks, and the fracture fluid permeates the rock extending the crack further, and further, and so on. Fractures are localized as pressure drops off with the rate of frictional loss, which is relevant to the distance from the well. Operators typically try to maintain "fracture width", or slow its decline following treatment, by introducing a proppant into the injected fluid – a material such as grains of sand, ceramic, or other particulate, thus preventing the fractures from closing when injection is stopped and pressure removed. Consideration of proppant strength and prevention of proppant failure becomes more important at greater depths where pressure and stresses on fractures are higher. The propped fracture is permeable enough to allow the flow of gas, oil, salt water and hydraulic fracturing fluids to the well.[45]

During the process, fracturing fluid leakoff (loss of fracturing fluid from the fracture channel into the surrounding permeable rock) occurs. If not controlled, it can exceed 70% of the injected volume. This may result in formation matrix damage, adverse formation fluid interaction, and altered fracture geometry, thereby decreasing efficiency.[46]

The location of one or more fractures along the length of the borehole is strictly controlled by various methods that create or seal holes in the side of the wellbore. Hydraulic fracturing is performed in cased wellbores, and the zones to be fractured are accessed by perforating the casing at those locations.[47]

Hydraulic-fracturing equipment used in oil and natural gas fields usually consists of a slurry blender, one or more high-pressure, high-volume fracturing pumps (typically powerful triplex or quintuplex pumps) and a monitoring unit. Associated equipment includes fracturing tanks, one or more units for storage and handling of proppant, high-pressure treating iron, a chemical additive unit (used to accurately monitor chemical addition), low-pressure flexible hoses, and many gauges and meters for flow rate, fluid density, and treating pressure.[48] Chemical additives are typically 0.5% percent of the total fluid volume. Fracturing equipment operates over a range of pressures and injection rates, and can reach up to 100 megapascals (15,000 psi) and 265 litres per second (9.4 cu ft/s) (100 barrels per minute).[49]

2.3.2 Well types

A distinction can be made between conventional, low-volume hydraulic fracturing, used to stimulate high-permeability reservoirs for a single well, and unconventional, high-volume hydraulic fracturing, used in the completion of tight gas and shale gas wells. High-volume hydraulic fracturing usually requires higher pressures than low-volume fracturing; the higher pressures are needed to push out larger volumes of fluid and proppant that extend farther from the borehole.[50]

Horizontal drilling involves wellbores with a terminal drillhole completed as a "lateral" that extends parallel with the rock layer containing the substance to be extracted. For example, laterals extend 1,500 to 5,000 feet (460 to 1,520 m) in the Barnett Shale basin in Texas, and up to 10,000 feet (3,000 m) in the Bakken formation in North Dakota. In contrast, a vertical well only accesses the thickness of the rock layer, typically 50–300 feet (15–91 m). Horizontal drilling reduces surface disruptions as fewer wells are required to access the same volume of rock.

Drilling often plugs up the pore spaces at the wellbore wall, reducing permeability at and near the wellbore. This reduces flow into the borehole from the surrounding rock formation, and partially seals off the borehole from the surrounding rock. Low-volume hydraulic fracturing can be used to restore permeability.[51]

2.3.3 Fracturing fluids

Main articles: Hydraulic fracturing proppants and List of additives for hydraulic fracturing

The main purposes of fracturing fluid are to extend fractures, add lubrication, change gel strength, and to carry proppant into the formation. There are two methods of transporting proppant in the fluid – high-rate and high-viscosity. High-viscosity fracturing tends to cause large dominant fractures, while high-rate (slickwater) fracturing causes small spread-out micro-fractures.

Water-soluble gelling agents (such as guar gum) increase viscosity and efficiently deliver proppant into the formation.[52]

Fluid is typically a slurry of water, proppant, and chemical additives.[53] Additionally, gels, foams, and compressed gases, including nitrogen, carbon dioxide and air can be injected. Typically, 90% of the fluid is water and 9.5% is sand with chemical additives accounting to about 0.5%.[45][54][55] However, fracturing fluids have been developed using liquefied petroleum gas (LPG) and propane in which water is unnecessary.[56]

The proppant is a granular material that prevents the created fractures from closing after the fracturing treatment. Types of proppant include silica sand, resin-coated sand, bauxite, and man-made ceramics. The choice of proppant depends on the type of permeability or grain strength needed. In some formations, where the pressure is great enough to crush grains of natural silica sand, higher-strength proppants such as bauxite or ceramics may be used. The most commonly used proppant is silica sand, though proppants of uniform size and shape, such as a ceramic proppant, are believed to be more effective.[57]

The fracturing fluid varies depending on fracturing type desired, and the conditions of specific wells being fractured, and water characteristics. The fluid can be gel, foam, or slickwater-based. Fluid choices are tradeoffs: more viscous fluids, such as gels, are better at keeping proppant in suspension; while less-viscous and lower-friction fluids, such as slickwater,

Water tanks preparing for hydraulic fracturing

allow fluid to be pumped at higher rates, to create fractures farther out from the wellbore. Important material properties of the fluid include viscosity, pH, various rheological factors, and others.

The water brought in is mixed with sand and chemicals to create fracking fluid. Approximately 40,000 gallons of chemicals are used per fracturing.[60] A typical fracture treatment uses between 3 and 12 additive chemicals.[45] Although there may be unconventional fracturing fluids, typical chemical additives can include one or more of the following:

- Acids—hydrochloric acid or acetic acid is used in the pre-fracturing stage for cleaning the perforations and initiating fissure in the near-wellbore rock.[55]

- Sodium chloride (salt)—delays breakdown of gel polymer chains.[55]

- Polyacrylamide and other friction reducers decrease turbulence in fluid flow and pipe friction, thus allowing the pumps to pump at a higher rate without having greater pressure on the surface.[55]

- Ethylene glycol—prevents formation of scale deposits in the pipe.[55]

- Borate salts—used for maintaining fluid viscosity during the temperature increase.[55]

- Sodium and potassium carbonates—used for maintaining effectiveness of crosslinkers.[55]

- Glutaraldehyde—used as disinfectant of the water (bacteria elimination).[55]

- Guar gum and other water-soluble gelling agents—increases viscosity of the fracturing fluid to deliver proppant into the formation more efficiently.[52][55]

Process of mixing water with hydraulic fracturing fluids to be injected into the ground

- Citric acid—used for corrosion prevention.

- Isopropanol—used to winterize the chemicals to ensure it doesn't freeze.[55]

The most common chemical used for hydraulic fracturing in the United States in 2005–2009 was methanol, while some other most widely used chemicals were isopropyl alcohol, 2-butoxyethanol, and ethylene glycol.[61]

Typical fluid types are:

- Conventional linear gels. These gels are cellulose derivative (carboxymethyl cellulose, hydroxyethyl cellulose, carboxymethyl hydroxyethyl cellulose, hydroxypropyl cellulose, hydroxyethyl methyl cellulose), guar or its derivatives (hydroxypropyl guar, carboxymethyl hydroxypropyl guar), mixed with other chemicals.

- Borate-crosslinked fluids. These are guar-based fluids cross-linked with boron ions (from aqueous borax/boric acid solution). These gels have higher viscosity at pH 9 onwards and are used to carry proppant. After the fracturing job, the pH is reduced to 3–4 so that the cross-links are broken, and the gel is less viscous and can be pumped out.

- Organometallic-crosslinked fluids zirconium, chromium, antimony, titanium salts are known to crosslink the guar-based gels. The crosslinking mechanism is not reversible, so once the proppant is pumped down along with cross-linked gel, the fracturing part is done. The gels are broken down with appropriate breakers.[52]

- Aluminium phosphate-ester oil gels. Aluminium phosphate and ester oils are slurried to form cross-linked gel. These are one of the first known gelling systems.

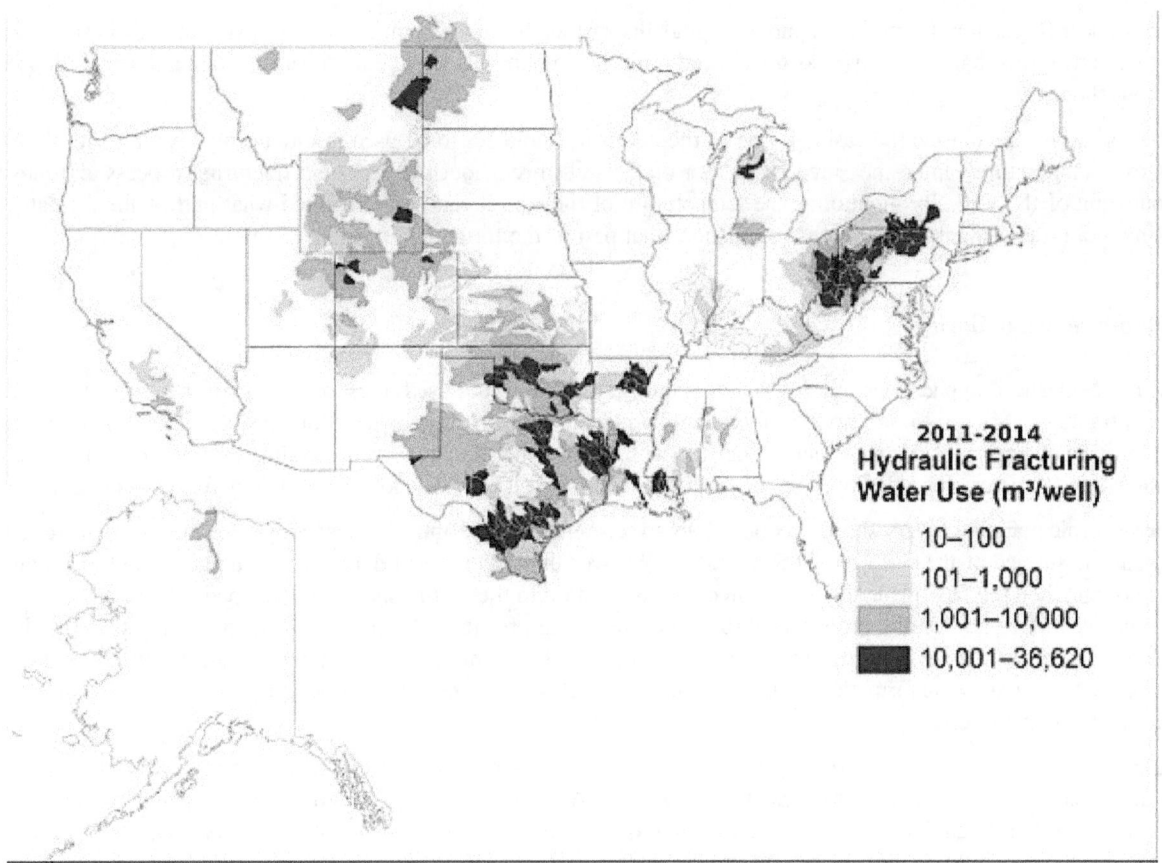

USGS map of water use from hydraulic fracturing between 2011 and 2014. One cubic meter of water is 264.172 gallons.[58][59]

For slickwater it is common to include sweeps or a temporary reduction in the proppant concentration to ensure the well is not overwhelmed with proppant causing a screen-off.[62] As the fracturing process proceeds, viscosity reducing agents such as oxidizers and enzyme breakers are sometimes then added to the fracturing fluid to deactivate the gelling agents and encourage flowback.[52] The oxidizer reacts with the gel to break it down, reducing the fluid's viscosity, and ensuring that no proppant is pulled from the formation. An enzyme acts as a catalyst for breaking down the gel. Sometimes pH modifiers are used to break down the crosslink at the end of a hydraulic fracturing job since many require a pH buffer system to stay viscous.[62] At the end of the job, the well is commonly flushed with water (sometimes blended with a friction reducing chemical) under pressure. Injected fluid is recovered to some degree and managed by several methods such as underground injection control, treatment and discharge, recycling, or temporary storage in pits or containers. New technology is continually being developed to better handle waste water and improve re-usability.[45]

2.3.4   Fracture monitoring

Measurements of the pressure and rate during the growth of a hydraulic fracture, with knowledge of fluid properties and proppant being injected into the well, provides the most common and simplest method of monitoring a hydraulic fracture treatment. This data along with knowledge of the underground geology can be used to model information such as length, width and conductivity of a propped fracture.[45]

Injection of radioactive tracers along with the fracturing fluid is sometimes used to determine the injection profile and location of created fractures.[63] Radiotracers are selected to have the readily detectable radiation, appropriate chemical properties, and a half life and toxicity level that will minimize initial and residual contamination.[64] Radioactive isotopes chemically bonded to glass (sand) and/or resin beads may also be injected to track fractures.[65] For example, plastic pellets coated with 10 GBq of Ag-110mm may be added to the proppant, or sand may be labelled with Ir-192, so that the proppant's progress can be monitored.[64] Radiotracers such as Tc-99m and I-131 are also used to measure flow rates.[64]

The Nuclear Regulatory Commission publishes guidelines which list a wide range of radioactive materials in solid, liquid and gaseous forms that may be used as tracers and limit the amount that may be used per injection and per well of each radionuclide.[65]

Fiber optics placed outside the casing is one of the newer technologies to be used in monitoring a well. With the fiber optics a temperature can be measured every foot of the well throughout the hydraulic fracturing process and into the production of the well. By monitoring the temperature of the well it can be determined what part of the formation is taking more fracturing fluid and during production what part of the formation is producing more.

Microseismic monitoring

For more advanced applications, microseismic monitoring is sometimes used to estimate the size and orientation of induced fractures. Microseismic activity is measured by placing an array of geophones in a nearby wellbore. By mapping the location of any small seismic events associated with the growing fracture, the approximate geometry of the fracture is inferred. Tiltmeter arrays deployed on the surface or down a well provide another technology for monitoring strain[66]

Microseismic mapping is very similar geophysically to seismology. In earthquake seismology, seismometers scattered on or near the surface of the earth record S-waves and P-waves that are released during an earthquake event. This allows for motion along the fault plane to be estimated and its location in the earth's subsurface mapped. Hydraulic fracturing, an increase in formation stress proportional to the net fracturing pressure, as well as an increase in pore pressure due to leakoff.[67] Tensile stresses are generated ahead of the fractures tip generating large amounts of shear stress. The increase in pore water pressure and formation stress combine and affect weaknesses (natural fractures, joints, and bedding planes) near the hydraulic fracture.[68]

Different methods have different location errors and advantages. Accuracy of microseismic event mapping is dependent on the signal-to-noise ratio and the distribution of sensors. Accuracy of events located by seismic inversion is improved by sensors placed in multiple azimuths from the monitored borehole. In a downhole array location, accuracy of events is improved by being close to the monitored borehole (high signal-to-noise ratio).

Monitoring of microseismic events induced by reservoir stimulation has become a key aspect in evaluation of hydraulic fractures, and their optimization. The main goal of hydraulic fracture monitoring is to completely characterize the induced fracture structure, and distribution of conductivity within a formation. Geomechanical analysis, such as understanding a formations material properties, in-situ conditions, and geometries, helps monitoring by providing a better definition of the environment in which the fracture network propagates.[69] The next task is to know the location of proppant within the fracture and the distribution of fracture conductivity. This can be monitored using multiple types of techniques to finally develop a reservoir model than accurately predicts well performance.

2.3.5   Horizontal completions

Since the early 2000s, advances in drilling and completion technology have made horizontal wellbores much more economical. Horizontal wellbores allow far greater exposure to a formation than conventional vertical wellbores. This is particularly useful in shale formations which do not have sufficient permeability to produce economically with a vertical well. Such wells, when drilled onshore, are now usually hydraulically fractured in a number of stages, especially in North America. The type of wellbore completion is used to determine how many times a formation is fractured, and at what locations along the horizontal section.[70]

In North America, shale reservoirs such as the Bakken, Barnett, Montney, Haynesville, Marcellus, and most recently the Eagle Ford, Niobrara and Utica shales are drilled horizontally through the producing interval(s), completed and fractured. The method by which the fractures are placed along the wellbore is most commonly achieved by one of two methods, known as "plug and perf" and "sliding sleeve".[71]

The wellbore for a plug-and-perf job is generally composed of standard steel casing, cemented or uncemented, set in the drilled hole. Once the drilling rig has been removed, a wireline truck is used to perforate near the bottom of the well, and then fracturing fluid is pumped. Then the wireline truck sets a plug in the well to temporarily seal off that section so the next section of the wellbore can be treated. Another stage is pumped, and the process is repeated along the horizontal length of the wellbore.[72]

The wellbore for the sliding sleeve technique is different in that the sliding sleeves are included at set spacings in the steel casing at the time it is set in place. The sliding sleeves are usually all closed at this time. When the well is due to be fractured, the bottom sliding sleeve is opened using one of several activation techniques and the first stage gets pumped. Once finished, the next sleeve is opened, concurrently isolating the previous stage, and the process repeats. For the sliding sleeve method, wireline is usually not required.

Sleeves

These completion techniques may allow for more than 30 stages to be pumped into the horizontal section of a single well if required, which is far more than would typically be pumped into a vertical well that had far fewer feet of producing zone exposed.[73]

2.4   Uses

Hydraulic fracturing is used to increase the rate at which fluids, such as petroleum, water, or natural gas can be recovered from subterranean natural reservoirs. Reservoirs are typically porous sandstones, limestones or dolomite rocks, but also include "unconventional reservoirs" such as shale rock or coal beds. Hydraulic fracturing enables the extraction of natural gas and oil from rock formations deep below the earth's surface (generally 2,000–6,000 m (5,000–20,000 ft)), which is greatly below typical groundwater reservoir levels. At such depth, there may be insufficient permeability or reservoir pressure to allow natural gas and oil to flow from the rock into the wellbore at high economic return. Thus, creating conductive fractures in the rock is instrumental in extraction from naturally impermeable shale reservoirs. Permeability is measured in the microdarcy to nanodarcy range.[74] Fractures are a conductive path connecting a larger volume of reservoir to the well. So-called "super fracking," creates cracks deeper in the rock formation to release more oil and gas, and increases efficiency.[75] The yield for typical shale bores generally falls off after the first year or two, but the peak producing life of a well can be extended to several decades.[76]

While the main industrial use of hydraulic fracturing is in stimulating production from oil and gas wells,[77][78][79] hydraulic fracturing is also applied:

- To stimulate groundwater wells[80]

- To precondition or induce rock cave-ins mining[81]

- As a means of enhancing waste remediation, usually hydrocarbon waste or spills[82]

- To dispose waste by injection deep into rock[83]

- To measure stress in the Earth[84]

- For electricity generation in enhanced geothermal systems[85]

- To increase injection rates for geologic sequestration of CO_2[86]

Since the late 1970s, hydraulic fracturing has been used, in some cases, to increase the yield of drinking water from wells in a number of countries, including the US, Australia, and South Africa.[87][88][89]

2.5 Economic effects

See also: Shale gas, Tight oil and Price of oil

Hydraulic fracturing has been seen as one of the key methods of extracting unconventional oil and unconventional gas resources. According to the International Energy Agency, the remaining technically recoverable resources of shale gas are estimated to amount to 208 trillion cubic metres (208,000 km^3), tight gas to 76 trillion cubic metres (76,000 km^3), and coalbed methane to 47 trillion cubic metres (47,000 km^3). As a rule, formations of these resources have lower permeability than conventional gas formations. Therefore, depending on the geological characteristics of the formation, specific technologies (such as hydraulic fracturing) are required. Although there are also other methods to extract these resources, such as conventional drilling or horizontal drilling, hydraulic fracturing is one of the key methods making their extraction economically viable. The multi-stage fracturing technique has facilitated the development of shale gas and light tight oil production in the United States and is believed to do so in the other countries with unconventional hydrocarbon resources.[6]

The National Petroleum Council estimates that hydraulic fracturing will eventually account for nearly 70% of natural gas development in North America.[90] Hydraulic fracturing and horizontal drilling apply the latest technologies and make it commercially viable to recover shale gas and oil. In the United States, 45% of domestic natural gas production and 17% of oil production would be lost within 5 years without usage of hydraulic fracturing.[91]

U.S.-based refineries have gained a competitive edge with their access to relatively inexpensive shale oil and Canadian crude. The U.S. is exporting more refined petroleum products, and also more liquified petroleum gas (LP gas). LP gas is produced from hydrocarbons called natural gas liquids, released by the hydraulic fracturing of petroliferous shale, in a variety of shale gas that's relatively easy to export. Propane, for example, costs around $620 a ton in the U.S. compared with more than $1,000 a ton in China, as of early 2014. Japan, for instance, is importing extra LP gas to fuel power plants, replacing idled nuclear plants. Trafigura Beheer BV, the third-largest independent trader of crude oil and refined products, said at the start of 2014 that "growth in U.S. shale production has turned the distillates market on its head."[92]

Some studies call into question the claim that what has been called the "shale gas revolution" has a significant macro-economic impact. A study released in the beginning of 2014 by the IDDRI concluded the contrary. It states that, on the long-term as well as on the short-run, the "shale gas revolution" due to hydraulic fracturing in the United States has had very little impact on economic growth and competitiveness.[93] The same report concludes that in Europe, using hydraulic fracturing would have very little advantage in terms of competitiveness and energy security. Indeed, for the period 2030-2035, shale gas is estimated to cover 3 to 10% of EU projected energy demand, which is not enough to have a significant impact on energetic independence and competitiveness.[93]

Hydrofracked shale oil and gas has the potential to alter the geography of energy production in the US.[94][95] In the short run, in counties with hydrofracturing employment in the oil and gas sector more than doubled in the last 10 years, with spill-overs in local transport-, construction but also manufacturing sectors.[94] The manufacturing sector benefits from lower energy prices, giving the US manufacturing sector a competitive edge. On average, natural gas prices have decreased by more than 30% in counties above shale deposits compared to the rest of the US. Some research has highlighted the negative effects on house prices for properties in the direct vicinity of fracturing wells.[96] Local house prices in Pennsylvania decrease if the property is close to a hydrofracking gas well and is not connected to city water, suggesting that the concerns of ground water pollution are priced by markets.

2.6   Public debate

Poster against hydraulic fracturing in Vitoria-Gasteiz, Spain, October 2012

2.6.1   Politics and public policy

An anti-fracking movement has emerged both internationally with involvement of international environmental organizations and nations such as France and locally in affected areas such as Balcombe in Sussex where the Balcombe drilling protest was in progress during mid-2013.[97] The considerable opposition against hydraulic fracturing activities in local townships in the United States has led companies to adopt a variety of public relations measures to reassure the public, including the employment of former military personnel with training in psychological warfare operations. According to Matt Pitzarella, the communications director at Range Resources, employees trained in the Middle East have been valuable to Range Resources in Pennsylvania, when dealing with emotionally charged township meetings and advising townships on zoning and local ordinances dealing with hydraulic fracturing.[98][99]

Protests have occasionally been marred by acts of violence. In March 2013, ten people were arrested[100] during an anti-fracking protest near New Matamoras, Ohio, after they illegally entered a development zone and latched themselves to drilling equipment. In northwest Pennsylvania, there was a drive-by shooting at a well site, in which someone shot two rounds of a small-caliber rifle in the direction of a drilling rig, before shouting profanities at the site and fleeing the scene.[101] In Washington County, Pennsylvania, a contractor working on a gas pipeline found a pipe bomb that had been placed where a pipeline was to be constructed, which local authorities said would have caused a "catastrophe" had they not discovered and detonated it.[102]

In 2014 a number of European officials provided circumstantial evidence that Eastern European protests against fracking may be sponsored by Gazprom, Russia's state-controlled gas company. Russian officials have on numerous occasions warned Europe that fracking "poses a huge environmental problem".[103]

2.6.2 Documentary films

Josh Fox's 2010 Academy Award nominated film *Gasland*[104] became a center of opposition to hydraulic fracturing of shale. The movie presented problems with ground water contamination near well sites in Pennsylvania, Wyoming, and Colorado.[105] *Energy in Depth*, an oil and gas industry lobbying group, called the film's facts into question.[106] In response, a rebuttal of *Energy in Depth*'s claims of inaccuracy was posted on *Gasland's* website.[107]

The Director of the Colorado Oil and Gas Conservation Commission (COGCC) offered to be interviewed as part of the film if he could review what was included from the interview in the final film but Fox declined the offer.[108] Exxon Mobil, Chevron Corporation and ConocoPhillips aired advertisements during 2011 and 2012 that claimed to describe the economic and environmental benefits of natural gas and argue that hydraulic fracturing was safe.[109]

The film *Promised Land*, starring Matt Damon, takes on hydraulic fracturing.[110] The gas industry is making plans to try to counter the film's criticisms of hydraulic fracturing with informational flyers, and Twitter and Facebook posts.[109]

On January 22, 2013 Northern Irish journalist and filmmaker Phelim McAleer released a crowdfunded[111] documentary called *FrackNation* as a response to the statements made by Fox in *Gasland*. *FrackNation* premiered on Mark Cuban's AXS TV. The premiere corresponded with the release of *Promised Land*.[112]

On April 21, 2013, Josh Fox released *Gasland 2*, a documentary that states that the gas industry's portrayal of natural gas as a clean and safe alternative to oil is a myth, and that hydraulically fractured wells inevitably leak over time, contaminating water and air, hurting families, and endangering the earth's climate with the potent greenhouse gas methane.

In 2014, Vido Innovations released the documentary *The Ethics of Fracking*. The film covers the politics, spiritual, scientific, medical and professional points of view on hydraulic fracturing. It also digs into the way the gas industry portrays fracking in their advertising.[113]

In 2015, the Canadian documentary film *Fractured Land* had its world premiere at the Hot Docs Canadian International Documentary Festival.[114]

2.6.3 Research issues

Typically the funding source of the research studies is a focal point of controversy. Concerns have been raised about research funded by foundations and corporations, or by environmental groups, which can at times lead to at least the appearance of unreliable studies.[115][116] Several organizations, researchers, and media outlets have reported difficulty in conducting and reporting the results of studies on hydraulic fracturing due to industry[117] and governmental pressure,[12] and expressed concern over possible censoring of environmental reports.[117][118][119] There is a need for more research into the environmental and health effects of the technique.[120][121][122][123]

2.7 Health risks

There is concern over the possible adverse public health implications of hydraulic fracturing activity.[120] A 2013 review on shale gas production in the United States stated, "with increasing numbers of drilling sites, more people are at risk from accidents and exposure to harmful substances used at fractured wells."[124] A 2011 hazard assessment recommended full disclosure of chemicals used for hydraulic fracturing and drilling as many have immediate health effects, and many may have long-term health effects.[125]

In June 2014 Public Health England published a review of the potential public health impacts of exposures to chemical and radioactive pollutants as a result of shale gas extraction in the UK, based on the examination of literature and data from countries where hydraulic fracturing already occurs.[121] The executive summary of the report stated: "An assessment of the currently available evidence indicates that the potential risks to public health from exposure to the emissions associated with shale gas extraction will be low if the operations are properly run and regulated. Most evidence suggests that contamination of groundwater, if it occurs, is most likely to be caused by leakage through the vertical borehole. Contamination of groundwater from the underground hydraulic fracturing process itself (ie the fracturing of the shale) is unlikely. However, surface spills of hydraulic fracturing fluids or wastewater may affect groundwater, and emissions to air also have the potential to impact on health. Where potential risks have been identified in the literature, the reported problems are typically a result of operational failure and a poor regulatory environment."[126]:iii

A 2012 report prepared for the European Union Directorate-General for the Environment identified potential risks to humans from air pollution and ground water contamination posed by hydraulic fracturing.[127] This led to a series of recommendations in 2014 to mitigate these concerns.[128][129] A 2012 guidance for pediatric nurses in the US said that hydraulic fracturing had a potential negative impact on public health and that pediatric nurses should be prepared to gather information on such topics so as to advocate for improved community health.[130]

2.8 Environmental impacts

Main article: Environmental impact of hydraulic fracturing
See also: Environmental impact of hydraulic fracturing in the United States and Exemptions for hydraulic fracturing under United States federal law

The environmental impacts of hydraulic fracturing include air emissions and climate change, high water consumption, water contamination, land use, risk of earthquakes, noise pollution, and health effects on humans. Air emissions are primarily methane that escapes from wells, along with industrial emissions from equipment used in the extraction process.[127] Modern UK and EU regulation requires zero emissions of methane, a potent greenhouse gas.[131] Escape of methane is a bigger problem in older wells than in ones built under more recent EU legislation.[127]

Hydraulic fracturing uses between 1.2 and 3.5 million US gallons (4,500 and 13,200 m^3) of water per well, with large projects using up to 5 million US gallons (19,000 m^3). Additional water is used when wells are refractured.[52][132] An average well requires 3 to 8 million US gallons (11,000 to 30,000 m^3) of water over its lifetime.[45] According to the Oxford Institute for Energy Studies, greater volumes of fracturing fluids are required in Europe, where the shale depths average 1.5 times greater than in the U.S.[133] Surface water may be contaminated through spillage and improperly built and maintained waste pits,[134] and ground water can be contaminated if the fluid is able to escape the formation being fractured (through, for example, abandoned wells) or by produced water (the returning fluids, which also contain dissolved constituents such as minerals and brine waters).[121] Produced water is managed by underground injection, municipal and commercial wastewater treatment and discharge, self-contained systems at well sites or fields, and recycling to fracture future wells.[135] Typically less than half of the produced water used to fracture the formation is recovered.[136]

About 3.6 hectares (8.9 acres) of land is needed per each drill pad for surface installations. Well pad and supporting structure construction significantly fragments landscapes which likely has negative effects on wildlife.[137] These sites need to be remediated after wells are exhausted.[127] Each well pad (in average 10 wells per pad) needs during preparatory and hydraulic fracturing process about 800 to 2,500 days of noisy activity, which affect both residents and local wildlife. In addition, noise is created by continuous truck traffic (sand, etc.) needed in hydraulic fracturing.[127] Research is underway to determine if human health has been affected by air and water pollution, and rigorous following of safety procedures and regulation is required to avoid harm and to manage the risk of accidents that could cause harm.[121]

In July 2013, the US Federal Railroad Administration listed oil contamination by hydraulic fracturing chemicals as "a possible cause" of corrosion in oil tank cars.[138]

Hydraulic fracturing sometimes causes induced seismicity or earthquakes. The magnitude of these events is usually too small to be detected at the surface, although tremors attributed to fluid injection into disposal wells have been large enough to have often been felt by people, and to have caused property damage and possibly injuries.[10][139][140][141][142]

Microseismic events are often used to map the horizontal and vertical extent of the fracturing.[66] A better understanding of the geology of the area being fracked and used for injection wells can be helpful in mitigating the potential for significant seismic events.[143]

2.9 Regulations

See also: Hydraulic fracturing by country and Regulation of hydraulic fracturing

Countries using or considering use of hydraulic fracturing have implemented different regulations, including developing federal and regional legislation, and local zoning limitations.[144][145] In 2011, after public pressure France became the first nation to ban hydraulic fracturing, based on the precautionary principle as well as the principle of preventive and corrective action of environmental hazards.[13][14][146][147] The ban was upheld by an October 2013 ruling of the Constitutional Council.[148] Some other countries such as Scotland have placed a temporary moratorium on the practice due to public health concerns and strong public opposition.[149] Countries like the United Kingdom and South Africa have lifted their bans, choosing to focus on regulation instead of outright prohibition.[150][151] Germany has announced draft regulations that would allow using hydraulic fracturing for the exploitation of shale gas deposits with the exception of wetland areas.[152]

The European Union has adopted a recommendation for minimum principles for using high-volume hydraulic fracturing.[15] Its regulatory regime requires full disclosure of all additives.[153] In the United States, the Ground Water Protection Council launched FracFocus.org, an online voluntary disclosure database for hydraulic fracturing fluids funded by oil and gas trade groups and the U.S. Department of Energy.[154][155] Hydraulic fracturing is excluded from the Safe Drinking Water Act's underground injection control's regulation, except when diesel fuel is used. The EPA assures surveillance of the issuance of drilling permits when diesel fuel is employed.[156]

In 2012, Vermont became the first state in the United States to ban hydraulic fracturing. On December 17, 2014, New York became the second state to issue a complete ban on any hydraulic fracturing due to potential risks to human health and the environment.[157][158][159]

2.10 See also

- Directional drilling

- Environmental concerns with electricity generation

- Environmental impact of hydraulic fracturing

- Environmental impact of petroleum

- Environmental impact of the oil shale industry

- ExxonMobil Electrofrac

- Hydraulic fracturing by country

- Hydraulic fracturing in the United States

- Hydraulic fracturing in the United Kingdom

- In-situ leach

2.11   References

[1] Luca Gandossi, Joint Research Centre, Institute for Energy and Transport (2013). An overview of hydraulic fracturing and other formation stimulation technologies for shale gas production (pdf). *Scientific and Technical Research series.* (Report) (European Commission Luxembourg: Publications Office of the European Union). Retrieved 29 May 2015.

[2] King, George E (2012), *Hydraulic fracturing 101* (PDF), Society of Petroleum Engineers, Paper 152596

[3] Staff. "State by state maps of hydraulic fracturing in US.". Fractracker.org. Retrieved 19 October 2013.

[4] Charlez, Philippe A. (1997). *Rock Mechanics: Petroleum Applications.* Paris: Editions Technip. p. 239. ISBN 9782710805861. Retrieved 2012-05-14.

[5] Blundell D., (2005). "Processes of tectonism, magmatism and mineralization: Lessons from Europe". *Ore Geology Reviews* **27**: 340.

[6] IEA (29 May 2012). *Golden Rules for a Golden Age of Gas. World Energy Outlook Special Report on Unconventional Gas* (PDF). OECD. pp. 18–27.

[7] Hillard Huntington et al. EMF 26: Changing the Game? Emissions and Market Implications of New Natural Gas Supplies Report. Stanford University. Energy Modeling Forum, 2013.

[8] Brown, Valerie J. (February 2007). "Industry Issues: Putting the Heat on Gas". *Environmental Health Perspectives* (US National Institute of Environmental Health Sciences) **115** (2): A76. doi:10.1289/ehp.115-a76. PMC 1817691. PMID 17384744. Retrieved 2012-05-01.

[9] V. J. Brown (February 2014). "Radionuclides in Fracking Wastewater: Managing a Toxic Blend". 122 (2). Environmental Health Perspectives. p. A50. Retrieved 27 May 2015.

[10] Kim, Won-Young 'Induced seismicity associated with fluid injection into a deep well in Youngstown, Ohio', Journal of Geophysical Research-Solid Earth

[11] US Geological Survey, Produced water, overview, accessed 8 Nov. 2014.

[12] Jared Metzker (7 August 2013). "Govt, Energy Industry Accused of Suppressing Fracking Dangers". Inter Press Service. Retrieved 28 December 2013.

[13] Patel, Tara (31 March 2011). "The French Public Says No to *'Le Fracking'*". *Bloomberg Businessweek.* Retrieved 22 February 2012.

[14] Patel, Tara (4 October 2011). "France to Keep Fracking Ban to Protect Environment, Sarkozy Says". *Bloomberg Businessweek.* Retrieved 22 February 2012.

[15] "Commission recommendation on minimum principles for the exploration and production of hydrocarbons (such as shale gas) using high-volume hydraulic fracturing (2014/70/EU)". *Official Journal of the European Union.* 22 January 2014. Retrieved 13 March 2014.

[16] Fjaer, E. (2008). "Mechanics of hydraulic fracturing". *Petroleum related rock mechanics.* Developments in petroleum science (2nd ed.). Elsevier. p. 369. ISBN 978-0-444-50260-5. Retrieved 2012-05-14.

[17] Price, N. J.; Cosgrove, J. W. (1990). *Analysis of geological structures.* Cambridge University Press. pp. 30–33. ISBN 978-0-521-31958-4. Retrieved 5 November 2011.

[18] Manthei, G.; Eisenblätter, J.; Kamlot, P. (2003). "Stress measurement in salt mines using a special hydraulic fracturing borehole tool". In Natau, Fecker & Pimentel. *Geotechnical Measurements and Modelling* (PDF). pp. 355–360. ISBN 90-5809-603-3. Retrieved 6 March 2012.

[19] Zoback, M.D. (2007). *Reservoir geomechanics.* Cambridge University Press. p. 18. ISBN 9780521146197. Retrieved 6 March 2012.

[20] Laubach, S. E.; Reed, R. M.; Olson, J. E.; Lander, R. H.; Bonnell, L. M. (2004). "Coevolution of crack-seal texture and fracture porosity in sedimentary rocks: cathodoluminescence observations of regional fractures". *Journal of Structural Geology* (Elsevier) **26** (5): 967–982. Bibcode:2004JSG....26..967L. doi:10.1016/j.jsg.2003.08.019. Retrieved 5 November 2011.

[21] Sibson, R. H.; Moore, J.; Rankin, A. H. (1975). "Seismic pumping--a hydrothermal fluid transport mechanism". *Journal of the Geological Society* (London: Geological Society) **131** (6): 653–659. doi:10.1144/gsjgs.131.6.0653. (subscription required). Retrieved 5 November 2011.

[22] Gill, R. (2010). *Igneous rocks and processes: a practical guide*. John Wiley and Sons. p. 102. ISBN 978-1-4443-3065-6. Retrieved 5 November 2011.

[23] "Shooters – A "Fracking" History". American Oil & Gas Historical Society. Retrieved 12 October 2014.

[24] "Acid fracturing". Society of Petroleum Engineers. Retrieved 12 October 2014.

[25] Montgomery, Carl T.; Smith, Michael B. (December 2010). "Hydraulic fracturing. History of an enduring technology" (PDF). *JPT Online* (Society of Petroleum Engineers): 26–41. Retrieved 13 May 2012.

[26] Energy Institute (February 2012). Fact-Based Regulation for Environmental Protection in Shale Gas Development (PDF) (Report). University of Texas at Austin. Retrieved 29 February 2012.

[27] A. J. Stark, A. Settari, J. R. Jones, Analysis of Hydraulic Fracturing of High Permeability Gas Wells to Reduce Non-darcy Skin Effects, Petroleum Society of Canada, Annual Technical Meeting, Jun 8 - 10, 1998, Calgary, Alberta.

[28] Mader, Detlef (1989). *Hydraulic Proppant Fracturing and Gravel Packing*. Elsevier. pp. 173–174; 202. ISBN 9780444873521.

[29] Ben E. Law and Charles W. Spencer, 1993, "Gas in tight reservoirs-an emerging major source of energy," *in* David G. Howell (ed.), *The Future of Energy Gasses*, US Geological Survey, Professional Paper 1570, p.233-252.

[30] C.R. Fast, G.B. Holman, and R. J. Covlin, "The application of massive hydraulic fracturing to the tight Muddy 'J' Formation, Wattenberg Field, Colorado," *in* Harry K. Veal, (ed.), *Exploration Frontiers of the Central and Southern Rockies* (Denver: Rocky Mountain Association of Geologists, 1977) 293-300.

[31] Robert Chancellor, "Mesaverde hydraulic fracture stimulation, northern Piceance Basin - progress report," *in* Harry K. Veal, (ed.), *Exploration Frontiers of the Central and Southern Rockies* (Denver: Rocky Mountain Association of Geologists, 1977) 285-291.

[32] C.E Bell and others, Effective diverting in horizontal wells in the Austin Chalk, Society of Petroleum Engineers conference paper, 1993.

[33] Robbins K. (2013). Awakening the Slumbering Giant: How Horizontal Drilling Technology Brought the Endangered Species Act to Bear on Hydraulic Fracturing. *Case Western Reserve Law Review*.

[34] E. O. Ray, Shale development in eastern Kentucky, US Energy Research and Development Administration, 1976.

[35] US Dept. of Energy, How is shale gas produced?, Apr. 2013.

[36] United States National Research Council, Committee to Review the Gas Research Institute's Research, Development and Demonstration Program, Gas Research Institute (1989). *A review of the management of the Gas Research Institute*. National Academies. p. ?.

[37] "US Government Role in Shale Gas Fracking: An Overview"

[38] *SPE production & operations* **20**. Society of Petroleum Engineers. 2005. p. 87.

[39] The Breakthrough Institute. Interview with Dan Steward, former Mitchell Energy Vice President. December 2011.

[40] Zuckerman, Gregory. "How fracking billionaires built their empires". *Quartz*. The Atlantic Media Company. Retrieved 15 November 2013.

[41] Wasley, Andrew (1 March 2013) On the frontline of Poland's fracking rush The Guardian, Retrieved 3 March 2013

[42] (7 August 2012) JKX Awards Fracking Contract for Ukrainian Prospect Natural Gas Europe, Retrieved 3 March 2013

[43] (18 February 2013) Turkey's shale gas hopes draw growing interest Reuters, Retrieved 3 March 2013

[44] "Hydraulic fracturing research study" (PDF). EPA. June 2010. EPA/600/F-10/002. Retrieved 2012-12-26.

[45] Ground Water Protection Council; ALL Consulting (April 2009). Modern Shale Gas Development in the United States: A Primer (PDF) (Report). DOE Office of Fossil Energy and National Energy Technology Laboratory. pp. 56–66. DE-FG26-04NT15455. Retrieved 24 February 2012.

[46] Penny, Glenn S.; Conway, Michael W.; Lee, Wellington (June 1985). "Control and Modeling of Fluid Leakoff During Hydraulic Fracturing". *Journal of Petroleum Technology* (Society of Petroleum Engineers) **37** (6): 1071–1081. doi:10.2118/12486-PA. Retrieved 2012-05-10.

[47] Arthur, J. Daniel; Bohm, Brian; Coughlin, Bobbi Jo; Layne, Mark (2008). Hydraulic Fracturing Considerations for Natural Gas Wells of the Fayetteville Shale (PDF) (Report). ALL Consulting. p. 10. Retrieved 2012-05-07.

[48] Chilingar, George V.; Robertson, John O.; Kumar, Sanjay (1989). *Surface Operations in Petroleum Production* **2**. Elsevier. pp. 143–152. ISBN 9780444426772.

[49] Love, Adam H. (December 2005). "Fracking: The Controversy Over its Safety for the Environment". Johnson Wright, Inc. Retrieved 2012-06-10.

[50] "Hydraulic Fracturing". University of Colorado Law School. Retrieved 2 June 2012.

[51] Wan Renpu (2011). *Advanced Well Completion Engineering*. Gulf Professional Publishing. p. 424. ISBN 9780123858689.

[52] Andrews, Anthony; et al. (30 October 2009). Unconventional Gas Shales: Development, Technology, and Policy Issues (PDF) (Report). Congressional Research Service. pp. 7; 23. Retrieved 22 February 2012.

[53] Ram Narayan (August 8, 2012). "From Food to Fracking: Guar Gum and International Regulation". *RegBlog*. University of Pennsylvania Law School. Retrieved 15 August 2012.

[54] Hartnett-White, K. (2011). "The Fracas About Fracking- Low Risk, High Reward, but the EPA is Against it" (PDF). National Review Online. Retrieved 2012-05-07.

[55] "Freeing Up Energy. Hydraulic Fracturing: Unlocking America's Natural Gas Resources" (PDF). American Petroleum Institute. 2010-07-19. Retrieved 2012-12-29.

[56] Brainard, Curtis (June 2013). "The Future of Energy". *Popular Science Magazine*. p. 59. Retrieved 1 January 2014.

[57] "CARBO ceramics". Retrieved 2011.

[58] "http://www.usgs.gov/newsroom/images/2015_06_30/water_use_for_fracking.jpg". *www.usgs.gov*. Retrieved 2015-07-03.

[59] Central, Bobby. "Water Use Rises as Fracking Expands". Retrieved 2015-07-03.

[60] Dong, Linda. "What goes in and out of Hydraulic Fracturing". *Dangers of Fracking*. Retrieved 2015.

[61] Chemicals Used in Hydraulic Fracturing (PDF) (Report). Committee on Energy and Commerce U.S. House of Representatives. April 18, 2011. p. ?.

[62] ALL Consulting (June 2012). The Modern Practices of Hydraulic Fracturing: A Focus on Canadian Resources (PDF) (Report). Canadian Association of Petroleum Producers. Retrieved 2012-08-04.

[63] Reis, John C. (1976). *Environmental Control in Petroleum Engineering*. Gulf Professional Publishers.

[64] Radiation Protection and the Management of Radioactive Waste in the Oil and Gas Industry (PDF) (Report). International Atomic Energy Agency. 2003. pp. 39–40. Retrieved 20 May 2012. Beta emitters, including ^{3}H and ^{14}C, may be used when it is feasible to use sampling techniques to detect the presence of the radiotracer, or when changes in activity concentration can be used as indicators of the properties of interest in the system. Gamma emitters, such as ^{46}Sc, ^{140}La, ^{56}Mn, ^{24}Na, ^{124}Sb, ^{192}Ir, ^{99}Tc^m, ^{131}I, ^{110}Ag^m, ^{41}Ar and ^{133}Xe are used extensively because of the ease with which they can be identified and measured. ... In order to aid the detection of any spillage of solutions of the 'soft' beta emitters, they are sometimes spiked with a short half-life gamma emitter such as ^{82}Br

[65] Jack E. Whitten, Steven R. Courtemanche, Andrea R. Jones, Richard E. Penrod, and David B. Fogl (Division of Industrial and Medical Nuclear Safety, Office of Nuclear Material Safety and Safeguards) (June 2000). "Consolidated Guidance About Materials Licenses: Program-Specific Guidance About Well Logging, Tracer, and Field Flood Study Licenses (NUREG-1556, Volume 14)". US Nuclear Regulatory Commission. Retrieved 19 April 2012. labeled Frac Sand...Sc-46, Br-82, Ag-110m, Sb-124, Ir-192

[66] Bennet, Les; et al. "The Source for Hydraulic Fracture Characterization" (PDF). *Oilfield Review* (Schlumberger) (Winter 2005/2006): 42–57. Retrieved 2012-09-30.

[67] Fehler, Michael C. (1989). "Stress Control of seismicity patterns observed during hydraulic fracturing experiments at the Fenton Hill hot dry rock geothermal energy site, New Mexico". *International Journal of Rock Mechanics and Mining Sciences & Geomechanics Abstracts*. 3 **26**. doi:10.1016/0148-9062(89)91971-2. Retrieved 1 January 2014.

[68] Le Calvez, Joel (2007). "Real-time microseismic monitoring of hydraulic fracture treatment: A tool to improve completion and reservoir management". *SPE Hydraulic Fracturing Technology Conference*.

[69] Cipolla, Craig (2010). "Hydraulic Fracture Monitoring to Reservoir Simulation: Maximizing Value". *SPE Annual Technical Conference and Exhibition*. Retrieved 1 January 2014.

[70] Seale, Rocky (July–August 2007). "Open hole completion systems enables multi-stage fracturing and stimulation along horizontal wellbores" (PDF). *Drilling Contractor* (Fracturing stimulation ed.). Retrieved October 1, 2009.

[71] "Completion Technologies". EERC. Retrieved 2012-09-30.

[72] "Energy from Shale". 2011.

[73] Mooney, Chris (2011). "The Truth About Fracking". *Scientific American* **305** (305): 80–85. Bibcode:2011SciAm.305d..80M. doi:10.1038/scientificamerican1111-80.

[74] "The Barnett Shale" (PDF). North Keller Neighbors Together. Retrieved 2012-05-14.

[75] David Wethe (19 January 2012). "Like Fracking? You'll Love 'Super Fracking'". *Businessweek*. Retrieved 22 January 2012.

[76] "Production Decline of a Natural Gas Well Over Time". *Geology.com*. The Geology Society of America. 3 January 2012. Retrieved 4 March 2012.

[77] Economides, Michael J. (2000). *Reservoir stimulation*. J. Wiley. p. P-2. ISBN 9780471491927.

[78] Gidley, John L. (1989). *Recent Advances in Hydraulic Fracturing*. SPE Monograph **12**. SPE. p. ?. ISBN 9781555630201.

[79] Ching H. Yew (1997). *Mechanics of Hydraulic Fracturing*. Gulf Professional Publishing. p. ?. ISBN 9780884154747.

[80] Banks, David; Odling, N. E.; Skarphagen, H.; Rohr-Torp, E. (May 1996). "Permeability and stress in crystalline rocks". *Terra Nova* **8** (3): 223–235. doi:10.1111/j.1365-3121.1996.tb00751.x.

[81] Brown, Edwin Thomas (2007) [2003]. *Block Caving Geomechanics* (2nd ed.). Indooroopilly, Queensland: Julius Kruttschnitt Mineral Research Centre, UQ. ISBN 978-0-9803622-0-6. Retrieved 2012-05-14.

[82] Frank, U.; Barkley, N. (February 1995). "Soil Remediation: Application of Innovative and Standard Technologies". *Journal of Hazardous Materials* **40** (2): 191–201. doi:10.1016/0304-3894(94)00069-S. ISSN 0304-3894. lcontribution= ignored (help) (subscription required)

[83] Bell, Frederic Gladstone (2004). *Engineering Geology and Construction*. Taylor & Francis. p. 670. ISBN 9780415259392.

[84] Aamodt, R. Lee; Kuriyagawa, Michio (1983). "Measurement of Instantaneous Shut-In Pressure in Crystalline Rock". *Hydraulic fracturing stress measurements*. National Academies. p. 139.

[85] "Geothermal Technologies Program: How an Enhanced Geothermal System Works". eere.energy.gov. 2011-02-16. Retrieved 2011-11-02.

[86] Miller, Bruce G. (2005). *Coal Energy Systems*. Sustainable World Series. Academic Press. p. 380. ISBN 9780124974517.

[87] Waltz, James; Decker, Tim L (1981), "Hydro-fracturing offers many benefits", *Johnson Driller's Journal* (2nd quarter): 4–9

[88] Williamson, WH (1982), "The use of hydraulic techniques to improve the yield of bores in fractured rocks", *Groundwater in Fractured Rock*, Conference Series (5), Australian Water Resources Council

[89] Less, C; Andersen, N (Feb 1994), "Hydrofracture: state of the art in South Africa", *Applied Hydrogeology*: 59–63

[90] National Petroleum Council, *Prudent Development: Realizing the Potential of North America's Abundant Natural Gas and Oil Resources*, September 15, 2011.

[91] IHS Global Insight, *Measuring the Economic and Energy Impacts of Proposals to Regulate Hydraulic Fracturing*, 2009.

[92] Asian Refiners Get Squeezed by U.S. Energy Boom, Wall Street Journal, Jan. 1, 2014

[93] Spencer T, Sartor O, Mathieu M "Unconventional wisdom: an economic analysis of US shale gas and implications for the EU", IDDRI, Paris, France, February 2014

[94] Fracking Growth - Estimating the Economic Impact of Shale Oil and Gas Development in the US, Fetzer, Thiemo (2014)

[95] Dutch Disease or Agglomeration? The Local Economic Effects of Natural Resource Booms in Modern America, Alcott and Kenniston (2013)

[96] The housing market impacts of shale gas development", by Lucija Muehlenbachs, Elisheba Spiller and Christopher Timmins. NBER Working Paper 19796, January 2014.

[97] Jan Goodey (1 August 2013). "The UK's anti fracking movement is growing". *The Ecologist.* Retrieved July 29, 2013.

[98] Javers, Eamon (8 Nov 2011). "Oil Executive: Military-Style 'Psy Ops' Experience Applied". *CNBC.*

[99] Phillips, Susan (9 Nov 2011). "'We're Dealing with an Insurgency,' says Energy Company Exec of Fracking Foes". *National Public Radio.*

[100] Palmer, Mike (27 March 2013). "Oil-gas boom spawns Harrison safety talks". *Times Leader.* Retrieved 27 March 2013.

[101] "Shots fired at W. Pa. gas drilling site". *Philadelphia Inquirer.* 12 March 2013. Retrieved 27 March 2013.

[102] Detrow, Scott (15 August 2012). "Pipe Bomb Found Near Allegheny County Pipeline". *NPR.* Retrieved 27 March 2013.

[103] Andrew Higgins (2014-11-30). "Russian Money Suspected Behind Fracking Protests". New York Times. Retrieved 2014-12-04.

[104] Documentary: *Gasland* (2010). 104 minutes.

[105] "Gasland". 2010. Retrieved 2012-05-14.

[106] "Gasland Debunked" (PDF). Energy in Depth. Retrieved 2012-05-14.

[107] "Affirming Gasland" (PDF). July 2010. Retrieved 2010-12-21.

[108] COGCC Gasland Correction Document *Colorado Department of Natural Resources* October 29, 2010

[109] Gilbert, Daniel (7 October 2012). "Matt Damon Fracking Film Lights Up Petroleum Lobby". *The Wall Street Journal* ((subscription required)). Retrieved 26 December 2012.

[110] Gerhardt, Tina (31 December 2012). "Matt Damon Exposes Fracking in Promised Land". *The Progressive.* Retrieved 4 January 2013.

[111] Kickstarter, FrackNation by Ann and Phelim Media LLC, April 6, 2012

[112] The Hollywood Reporter, Mark Cuban's AXS TV Picks Up Pro-Fracking Documentary 'FrackNation', December 17, 2012

[113] "The Ethics of Fracking". *Green Planet Films.*

[114] "'Fractured Land' Doc Coming to VIFF". *The Tyee.* 2015-09-09. Retrieved 2015-10-20.

[115] Deller, Steven; Schreiber, Andrew (2012). "Mining and Community Economic Growth" (PDF). *The Review of Regional Studies* **42**: 121–141. Retrieved 3 March 2013.

[116] Soraghan, Mike (12 March 2012). "Quiet foundation funds the 'anti-fracking' fight". *E&E News.* Retrieved 27 March 2013. In our work to oppose fracking, the Park Foundation has simply helped to fuel an army of courageous individuals and NGOs,' or non-governmental organizations, said Adelaide Park Gomer, foundation president and Park heir, in a speech late last year.

[117] Urbina, Ian (3 March 2011). "Pressure Limits Efforts to Police Drilling for Gas". *The New York Times.* Retrieved 23 February 2012. More than a quarter-century of efforts by some lawmakers and regulators to force the federal government to police the industry better have been thwarted, as E.P.A. studies have been repeatedly narrowed in scope and important findings have been removed

[118] "The Debate Over the Hydrofracking Study's Scope". *The New York Times.* 3 March 2011. Retrieved 1 May 2012. While environmentalists have aggressively lobbied the agency to broaden the scope of the study, industry has lobbied the agency to narrow this focus

[119] "Natural Gas Documents". *The New York Times.* 27 February 2011. Retrieved 5 May 2012. The Times reviewed more than 30,000 pages of documents obtained through open records requests of state and federal agencies and by visiting various regional offices that oversee drilling in Pennsylvania. Some of the documents were leaked by state or federal officials.

[120] Finkel ML, Hays J (October 2013). "The implications of unconventional drilling for natural gas: a global public health concern". *Public Health* (Review) **127** (10): 889–93. doi:10.1016/j.puhe.2013.07.005. PMID 24119661.

[121] Public Health England. 25 June 2014 PHE-CRCE-009: Review of the potential public health impacts of exposures to chemical and radioactive pollutants as a result of shale gas extraction ISBN 978-0-85951-752-2

[122] Drajem, Mark (11 January 2012). "Fracking Political Support Unshaken by Doctors' Call for Ban". Bloomberg. Retrieved 19 January 2012.

[123] Alex Wayne (4 January 2012). "Health Effects of Fracking Need Study, Says CDC Scientist". *Bloomberg Businessweek.* Retrieved 29 February 2012.

[124] Centner, Terence J. (September 2013). "Oversight of shale gas production in the United States and the disclosure of toxic substances". *Resources Policy* **38** (3): 233–240. doi:10.1016/j.resourpol.2013.03.001. Retrieved 29 September 2014.

[125] Colborn, Theo; et al. (September 20, 2011). "Natural Gas Operations from a Public Health Perspective" (PDF). *Human and Ecological Risk Assessment* **17** (5): 1039–1056. doi:10.1080/10807039.2011.605662.

[126] A. Kibble, T. Cabianca, Z. Daraktchieva, T. Gooding, J. Smithard, G. Kowalczyk, N. P. McColl, M. Singh, L. Mitchem, P. Lamb, S. Vardoulakis and R. Kamanyire (January 2014). Review of the Potential Public Health Impacts of Exposures to Chemical and Radioactive Pollutants as a Result of the Shale Gas Extraction Process (PDF) (Report). Public Health England. PHE-CRCE-009.

[127] Broomfield, Mark (2012-08-10). Support to the identification of potential risks for the environment and human health arising from hydrocarbons operations involving hydraulic fracturing in Europe (PDF) (Report) (17c). European Commission. pp. vi–xvi. ED57281. Retrieved 2014-09-29.

[128] "EU Commission minimum principles for the exploration and production of hydrocarbons (such as shale gas) using high-volume hydraulic fracturing". EUR LEX. Retrieved Nov 2014.

[129] "Energy and environment". EUR LEX.

[130] Lauver LS (August 2012). "Environmental health advocacy: an overview of natural gas drilling in northeast Pennsylvania and implications for pediatric nursing". *J Pediatr Nurs* **27** (4): 383–9. doi:10.1016/j.pedn.2011.07.012. PMID 22703686.

[131] "Air Quality" (PDF). DECC.

[132] Abdalla, Charles W.; Drohan, Joy R. (2010). Water Withdrawals for Development of Marcellus Shale Gas in Pennsylvania. Introduction to Pennsylvania's Water Resources (PDF) (Report). The Pennsylvania State University. Retrieved 16 September 2012. Hydrofracturing a horizontal Marcellus well may use 4 to 8 million gallons of water, typically within about 1 week. However, based on experiences in other major U.S. shale gas fields, some Marcellus wells may need to be hydrofractured several times over their productive life (typically five to twenty years or more)

[133] Faucon, Benoît (17 September 2012). "Shale-Gas Boom Hits Eastern Europe". WSJ.com. Retrieved 17 September 2012.

[134] New Research of Surface Spills in Fracking Industry. (2013). Professional Safety, 58(9), 18.

[135] Logan, Jeffrey (2012). Natural Gas and the Transformation of the U.S. Energy Sector: Electricity (PDF) (Report). Joint Institute for Strategic Energy Analysis. Retrieved 27 March 2013.

[136] Köster, Vera. "What is Shale Gas? How Does Fracking Work?". *www.chemistryviews.org.* Retrieved 4 December 2014.

[137] http://link.springer.com/article/10.1007/s00267-014-0440-6

[138] Frederick J. Herrmann, Federal Railroad Administration, letter to American Petroleum Institute, 17 July 2013, p.4.

[139] Zoback, Mark; Kitasei, Saya; Copithorne, Brad (July 2010). Addressing the Environmental Risks from Shale Gas Development (PDF) (Report). Worldwatch Institute. p. 9. Retrieved 2012-05-24.

[140] Begley, Sharon; McAllister, Edward (12 July 2013). "News in Science: Earthquakes may trigger fracking tremors". *ABC Science* (Reuters). Retrieved 17 December 2013.

[141] "Fracking tests near Blackpool 'likely cause' of tremors". *BBC News*. 2 November 2011. Retrieved 22 February 2012.

[142] Injection-Induced Earthquakes, Science 341, W. L. Ellsworth, 2013. Retrieved 31 December 2014.

[143] Managing the seismic risk posed by wastewater disposal, Earth Magazine, 57:38–43 (2012), M. D. Zoback. Retrieved 31 December 2014.

[144] Nolon, John R.; Polidoro, Victoria (2012). "Hydrofracking: Disturbances Both Geological and Political: Who Decides?" (PDF). *The Urban Lawyer* **44** (3): 1–14. Retrieved 2012-12-21.

[145] Negro, Sorrell E. (February 2012). "Fracking Wars: Federal, State, and Local Conflicts over the Regulation of Natural Gas Activities" (PDF). *Zoning and Planning Law Report* (Thomson Reuters) **35** (2): 1–14. Retrieved 2014-05-01.

[146] "LOI n° 2011-835 du 13 juillet 2011 visant à interdire l'exploration et l'exploitation des mines d'hydrocarbures liquides ou gazeux par fracturation hydraulique et à abroger les permis exclusifs de recherches comportant des projets ayant recours à cette technique"

[147] "Article L 110-1 du Code de l'Environnement"

[148] "Fracking ban upheld by French court". *BBC*. 11 October 2013. Retrieved 16 October 2013.

[149] Moore, Robbie. "Fracking, PR, and the Greening of Gas". *The International*. Retrieved 16 March 2013.

[150] Bakewell, Sally (13 December 2012). "U.K. Government Lifts Ban on Shale Gas Fracking". Bloomberg. Retrieved 26 March 2013.

[151] Hweshe, Francis (17 September 2012). "South Africa: International Groups Rally Against Fracking, TKAG Claims". *West Cape News*. Retrieved 11 February 2014.

[152] Nicola, Stefan; Andersen, Tino (26 February 2013). "Germany agrees on regulations to allow fracking for shale gas". Bloomberg. Retrieved 1 May 2014.

[153] Healy, Dave (July 2012). Hydraulic Fracturing or 'Fracking': A Short Summary of Current Knowledge and Potential Environmental Impacts (PDF) (Report). Environmental Protection Agency. Retrieved 28 July 2013.

[154] Hass, Benjamin (14 August 2012). "Fracking Hazards Obscured in Failure to Disclose Wells". Bloomberg. Retrieved 27 March 2013.

[155] Soraghan, Mike (13 December 2013). "White House official backs FracFocus as preferred disclosure method". *E&E News*. Retrieved 27 March 2013.

[156] , Environmental Protection Agency

[157] "Gov. Cuomo Makes Sense on Fracking". *The New York Times*. December 17, 2014. Retrieved December 18, 2014.

[158] Nearing, Brian (December 18, 2014). "Citing perils, state bans fracking". *Times Union*. Retrieved January 25, 2015.

[159] Brady, Jeff (December 18, 2014). "Citing Health, Environment Concerns, New York Moves To Ban Fracking". *NPR*. Retrieved January 25, 2015.

2.12   External links

- Natural Gas Extraction—Hydraulic Fracturing (EPA website)

- EPA's Draft Hydraulic Fracturing Study Plan

- The British Columbia (Canada) Oil and Gas Commission mandatory disclosure of hydraulic fracturing fluids

- Hydraulic Fracturing: Selected Legal Issues Congressional Research Service

- Fracking collected news and commentary at *ProPublica*

- Hydraulic Fracturing at Earthworks

- FracFocus Searchable database with chemical composition of fracking fluid of individual wells

- FracTracker.org: Maps, data, and articles from news, government, industry, and academic sources.

- 60 Minutes Report on Hydraulic Fracturing.

2.13 Further reading

- Kiparsky, Michael; Hein, Jayni Foley (April 2013). "Regulation of Hydraulic Fracturing in California: A Wastewater and Water Quality Perspective" (PDF). University of California Center for Law, Energy, and the Environment. Retrieved 2014-05-01.

- Ridlington, Elizabeth; John Rumpler (October 3, 2013). "Fracking by the numbers". *Environment America.*

- "DISH, TExas Exposure Investigation" (PDF). Texas DSHS. Retrieved 27 March 2013.

- de Pater, C.J.; Baisch, S. (2 November 2011). Geomechanical Study of Bowland Shale Seismicity (PDF) (Report). Cuadrilla Resources. Retrieved 22 February 2012.

- McKenzie, Lisa; Witter, Roxana; Newman, Lee; Adgate, John (2012). "Human health risk assessment of air emissions from development of unconventional natural gas resources". *Science of the Total Environment* **424**: 79–87. doi:10.1016/j.scitotenv.2012.02.018. PMID 22444058.

- "The Hydraulic Fracturing Water Cycle". EPA. 16 March 2014. Retrieved 2014-10-10.

- Fernandez, John Michael; Gunter, Matthew. "Hydraulic Fracturing: Environmentally Friendly Practices" (PDF). Houston Advanced Research Center. Retrieved 2012-12-29.

- Colborn, Theo; Kwiatkowski, Carol; Schultz, Kim; Bachran, Mary (2011). "Natural gas operations from public health perspective". *Human and Ecological Risk Assessment: an International Journal* **17** (5): 1039–1056. doi:10.1080/10807039.2011.605662.

- Abdalla, Charles W.; Drohan, Joy R.; Blunk, Kristen Saacke; Edson, Jessie (2014). Marcellus Shale Wastewater Issues in Pennsylvania—Current and Emerging Treatment and Disposal Technologies (PDF) (Report). Penn State Extension. Retrieved 2014-10-11.

- Arthur, J. Daniel; Langhus, Bruce; Alleman, David (2008). An overview of modern shale gas development in the United States (PDF) (Report). ALL Consulting. p. 21. Retrieved 2012-05-07.

- Howe, J. Cullen; Del Percio, Stephen. The Legal and Regulatory Landscape of Hydraulic Fracturing (Report). LexisNexis. Retrieved 2014-05-07.

- Molofsky, L. J.; Connor, J. A.; Shahla, K. F.; Wylie, A. S.; Wagner, T. (December 5, 2011). "Methane in Pennsylvania Water Wells Unrelated to Marcellus Shale Fracturing". *Oil and Gas Journal* (Pennwell Corporation) **109** (49): 54–67.

- IEA (2011). *World Energy Outlook 2011*. OECD. pp. 91; 164. ISBN 9789264124134.

- "How is hydraulic fracturing related to earthquakes and tremors?". USGS. Retrieved 4 November 2012.

- Moniz, Ernest J.; et al. (June 2011). The Future of Natural Gas: An Interdisciplinary MIT Study (PDF) (Report). Massachusetts Institute of Technology. Retrieved 1 June 2012.

- Biello, David (30 March 2010). "Natural gas cracked out of shale deposits may mean the U.S. has a stable supply for a century – but at what cost to the environment and human health?". Scientific American. Retrieved 23 March 2012.

- Schmidt, Charles (1 August 2011). "Blind Rush? Shale Gas Boom Proceeds Amid Human Health Questions". *Environmental Health Perspectives* **119** (8): a348–a353. doi:10.1289/ehp.119-a348. PMC 3237379. PMID 21807583.

- Allen, David T.; Torres, Vincent N.; Thomas, James; Sullivan, David W.; Harrison, Matthew; Hendler, Al; Herndon, Scott C.; Kolb, Charles E.; Fraser, Matthew P.; Hill, A. Daniel; Lamb, Brian K.; Miskimins, Jennifer; Sawyer, Robert F.; Seinfeld, John H. (16 September 2013). "Measurements of methane emissions at natural gas production sites in the United States" (PDF). *Proceedings of the National Academy of Sciences*. doi:10.1073/pnas.1304880110. Retrieved 2013-10-02.

- Kassotis, Christopher D.; Tillitt, Donald E.; Davis, J. Wade; Hormann, Annette M.; Nagel, Susan C. (March 2014). "Estrogen and Androgen Receptor Activities of Hydraulic Fracturing Chemicals and Surface and Ground Water in a Drilling-Dense Region". *Endocrinology* **155** (3). doi:10.1210/en.2013-1697. Retrieved 24 December 2013.

- Chalupka, S. (October 2012). "Occupational Silica Exposure in Hydraulic Fracturing". *Workplace Health & Safety* **60** (10): 460. doi:10.3928/21650799-20120926-70. Retrieved 10 October 2014.

- Smith, S. (1 August 2014). "Respirators Are Not Enough: New Study Examines Worker Exposure to Silica in Hydraulic Fracturing Operations". *EHS Today*. Retrieved 10 October 2014.

- "Waste water (flowback)from hydraulic fracturing" (PDF). Ohio Department of Natural Resources. Retrieved 29 June 2013.

- Spath, Ph.D., P.E., David P. (November 1997). Policy Memo 97-005 Policy Guidance for Direct Domestic Use of Extremely Impaired Sources (PDF) (Report). State of California Department of Health Services. Retrieved 7 October 2014.

- Weinhold, Bob (19 September 2012). "Unknown Quantity: Regulating Radionuclides in Tap Water". *Environmental Health Perspectives*. NIEHS, NIH. Retrieved 11 February 2012. Examples of human activities that may lead to radionuclide exposure include mining, milling, and processing of radioactive substances; wastewater releases from the hydraulic fracturing of oil and natural gas wells... Mining and hydraulic fracturing, or "fracking", can concentrate levels of uranium (as well as radium, radon, and thorium) in wastewater...

- Rachel Maddow, Terrence Henry (7 August 2012). *Rachel Maddow Show: Fracking waste messes with Texas* (video). MSNBC. Event occurs at 9:24 - 10:35.

- Cothren, Jackson. Modeling the Effects of Non-Riparian Surface Water Diversions on Flow Conditions in the Little Red Watershed (PDF) (Report). U. S. Geological Survey, Arkansas Water Science Center Arkansas Water Resources Center, American Water Resources Association, Arkansas State Section Fayetteville Shale Symposium 2012. p. 12. Retrieved 16 September 2012. ...each well requires between 3 and 7 million gallons of water for hydraulic fracturing and the number of wells is expected to grow in the future

- Janco, David F. (1 February 2007). PADEP Determination Letter No. 970. Diminution of Snow Shoe Borough Authority Water Well No. 2; primary water source for about 1,000 homes and businesses in and around the borough; contested by Range Resources. Determination Letter acquired by the Scranton Times-Tribune via Right-To-Know Law request. (PDF) (Report). Scranton Times-Tribune. Retrieved 27 December 2013.

- Janco, David F. (3 January 2008). PADEP Determination Letter No. 352 Determination Letter acquired by the Scranton Times-Tribune via Right-To-Know Law request. Order: Atlas Miller 42 and 43 gas wells; Aug 2007 investigation; supplied temporary buffalo for two springs, ordered to permanently replace supplies (PDF) (Report). Scranton Times-Tribune. Retrieved 27 December 2013.

- Lustgarten, Abrahm (21 June 2012). "Are Fracking Wastewater Wells Poisoning the Ground beneath Our Feet? Leaking injection wells may pose a risk--and the science has not kept pace with the growing glut of wastewater". *Scientific American*. Retrieved 2014-10-11.

- Rabinowitz, Peter M.; Rabinowitz, Ilya B.; Slizovskiy, Vanessa; Lamers, Sally J.; Trufan, Theodore R.; Holford, James D.; Dziura, Peter N.; Peduzzi, Michael J.; Kane, John S.; Reif, John; Weiss, Theresa R.; Stowe1, Meredith H. (2014). "Proximity to Natural Gas Wells and Reported Health Status: Results of a Household Survey in Washington County, Pennsylvania". *Environmental Health Perspectives* (US National Institute of Environmental Health Sciences). doi:10.1289/ehp.1307732. Retrieved 2014-10-07.

- Arthur, J. Daniel; Uretsky, Mike; Wilson, Preston (May 5–6, 2010). *Water Resources and Use for Hydraulic Fracturing in the Marcellus Shale Region* (PDF). Meeting of the American Institute of Professional Geologists. Pittsburgh: ALL Consulting. p. 3. Retrieved 2012-05-09.

- Colborn, Theo; Kwiatkowski, Carol; Schultz, Kim; Bachran, Mary (2011). "Natural Gas Operations from a Public Health Perspective" (PDF). *Human and Ecological Risk Assessment: an International Journal* (Taylor & Francis) **17** (5): 1039–1056. doi:10.1080/10807039.2011.605662.

- Osborn, Stephen G.; Vengosh, Avner; Warner, Nathaniel R.; Jackson, Robert B. (2011-05-17). "Methane contamination of drinking water accompanying gas-well drilling and hydraulic fracturing" (PDF). *Proceedings of the National Academy of Sciences of the United States of America* **108** (20): 8172–8176. doi:10.1073/pnas.1100682108. Retrieved 2011-10-14.

- Nicholas St. Fleur (19 December 2014). "The Alarming Research Behind New York's Fracking Ban -- an analysis of the findings in Governor Andrew Cuomo's 184-page review of hydraulic fracturing". *The Atlantic*. Retrieved 21 December 2014.

- Gallegos, T.J. and B.A. Varela (2015). Hydraulic Fracturing Distributions and Treatment Fluids, Additives, Proppants, and Water Volumes Applied to Wells Drilled in the United States from 1947 through 2010. U.S. Geological Survey.

Chapter 3

Regulation of hydraulic fracturing

Countries using or considering to use hydraulic fracturing have implemented different regulations, including developing federal and regional legislation, and local zoning limitations.[1][2] In 2011, after public pressure France became the first nation to ban hydraulic fracturing, based on the precautionary principle as well as the principal of preventive and corrective action of environmental hazards.[3][4][5][6] The ban was upheld by an October 2013 ruling of the Constitutional Council.[7] Some other countries have placed a temporary moratorium on the practice.[8] Countries like the United Kingdom and South Africa, have lifted their bans, choosing to focus on regulation instead of outright prohibition.[9][10] Germany has announced draft regulations that would allow using hydraulic fracturing for the exploitation of shale gas deposits with the exception of wetland areas.[11]

The European Union has adopted a recommendation for minimum principles for using high-volume hydraulic fracturing.[12] Its regulatory regime requires full disclosure of all additives.[13] In the United States, the Ground Water Protection Council launched FracFocus.org, an online voluntary disclosure database for hydraulic fracturing fluids funded by oil and gas trade groups and the U.S. Department of Energy.[14][15] Hydraulic fracturing is excluded from the Safe Drinking Water Act's underground injection control's regulation, except when diesel fuel is used. The EPA assures surveillance of the issuance of drilling permits when diesel fuel is employed.[16]

On 17 December 2014, New York state issued a statewide ban on hydraulic fracturing, becoming the second state in the United States to issue such a ban after Vermont.[17]

3.1 Approaches

3.1.1 Risk-based approach

The main tool used by this approach is risk assessment. A risk assessment method, based on experimenting and assessing risk ex-post, once the technology is in place. In the context of hydraulic fracturing, it means that drilling permits are issued and exploitation conducted before the potential risks on the environment and human health are known. The risk-based approach mainly relies on a discourse that sacralizes technological innovations as an intrinsic good, and the analysis of such innovations, such as hydraulic fracturing, is made on a sole cost-benefit framework, which does not allow prevention or ex-ante debates on the use of the technology.[18] This is also referred to as "learning-by-doing".[19] A risk assessment method has for instance led to regulations that exist in the hydraulic fracturing in the United States (EPA will release its study on the impact of hydraulic fracturing on groundwater in 2014, though hydraulic fracturing has been used for more than 60 years. Commissions that have been implemented in the US to regulate the use of hydraulic fracturing have been created after hydraulic fracturing had started in their area of regulation. This is for instance the case in the Marcellus shale area where three regulatory committees were implemented ex-post.[20]

Academic scholars who have studied the perception of hydraulic fracturing in the North of England have raised two main critiques of this approach. Firstly, it takes scientific issues out of the public debate since there is no debate on the use of a technology but on its impacts. Secondly, it does not prevent environmental harm from happening since risks are taken then assessed instead of evaluated then taken as it would be the case with a precautionary approach to scientific debates. The relevance and reliability of risk assessments in hydraulic fracturing communities has also been debated amongst environmental groups, health scientists, and industry leaders. A study has epitomized this point: the participants to regulatory committees of the Marcellus shale have, for a majority, raised concerns about public health although nobody in these regulatory committees had expertise in public health. That highlights a possible underestimation of public health risks due to hydraulic fracturing. Moreover, more than a quarter of the participants raised concerns about the neutrality of the regulatory committees given the important weigh of the hydraulic fracturing industry.[20] The risks, to some like the participants of the Marcellus Shale regulatory committees, are overplayed and the current research is insufficient in showing the link between hydraulic fracturing and adverse health effects, while to others like local environmental groups the risks are obvious and risk assessment is underfunded.[19]

3.1.2 Precaution-based approach

The second approach relies on the precautionary principle and the principal of preventive and corrective action of environmental hazards, using the best available techniques with an acceptable economic cost to insure the protection, the valuation, the restoration, management of spaces, resources and natural environments, of animal and vegetal species, of ecological diversity and equilibriums.[6] The precautionary approach has led to regulations as implemented in France and Vermont, banning hydraulic fracturing.[5][21]

Such an approach is called upon by social sciences and the public as studies have shown in the North of England and Australia.[18][19] Indeed, in Australia, the anthropologist who studied the use of hydraulic fracturing concluded that the risk-based approach was closing down the debate on the ethics of such a practice, therefore avoiding questions on broader concerns that merely the risks implied by hydraulic fracturing. In the North of England, levels of concerns registered in the deliberative focus groups studied were higher regarding the framing of the debate, meaning the fact that people did not have a voice in the energetic choices that were made, including the use of hydraulic fracturing. Concerns relative to risks of seismicity and health issues were also important to the public, but less than this. A reason for that is that being withdrawn the right to participate in the decision-making triggered opposition of both supporters and opponents of hydraulic fracturing.

The points made to defend such an approach often relate to climate change and the impact on the direct environment; related to public concerns on the rural landscape for instance in the UK.[19] Energetic choices indeed have an impact on climate change since greenhouse gas emissions from fossil fuels extraction such as shale gas and oil contribute to climate change. Therefore, people have in the UK raised concerns about the exploitation of these resources, not just hydraulic fracturing as a method. They would hence prefer a precaution-based approach to decide whether or not, regarding the issue of climate change, they want to exploit shale gas and oil.

3.2 Framing of the debate

There are two main areas of interest regarding how debates on hydraulic fracturing for the exploitation of unconventional oil and gas have been conducted.

3.2.1 "Learning-by-doing" and the displacement of ethics

A risk-based approach is often referred to as "learning-by-doing" by social sciences. Social sciences have raised two main critiques of this approach. Firstly, it takes scientific issues out of the public debate since there is no debate on the use of a technology but on its impacts. Secondly, it does not prevent environmental harm from happening since risks are taken then assessed instead of evaluated then taken. Public concerns are shown to be really linked to these issues of scientific approach. Indeed, the public in the North of England for instance fears "the denial of the deliberation of the values embedded in the development and application of that technology, as well as the future it is working towards" more than risks themselves. The legitimacy of the method is only questioned after its implementation, not before. This vision separates risks and impacts from the values entitled by a technology. For instance, hydraulic fracturing entitles a transitional fuel for its supporters whereas for its opponents it represents a fossil fuel exacerbating the greenhouse effect and global warming. Not asking these questions leads to seeing only the mere economic cost-benefit analysis.[19]

This is linked to a pattern of preventing non-experts from taking part in scientific-technological debates, including their ethical issues. An answer to that problem is seen to be increased public participation so as to have the public deciding which issues to address and what political and ethical norms to adopt as a society. Another public concern with the "learning-by-doing" approach is that the speed of innovation may exceed the speed of regulation and since innovation is seen as serving private interests, potentially at the expense of social good, it is a matter of public concern. Science and Technology Studies have theorized "slowing-down" and the precautionary principle as answers. The claim is that the possibility of an issue is legitimate and should be taken into account before any action is taken.[19]

3.2.2   Variations in risk-assessment of environmental impacts of hydraulic fracturing

Issues also exist regarding the way risk assessment is conducted and whether it reflects some interests more than others. Firstly, an issue exists about whether risk assessment authorities are able to judge the impact of hydraulic fracturing in public health. A study conducted on the advisory committees of the Marcellus Shale gas area[20] has shown that not a single member of these committees had public health expertise and that some concern existed about whether the commissions were not biased in their composition. Indeed, among 51 members of the committees, there is no evidence that a single one has any expertise in environmental public health, even after enlarging the category of experts to "include medical and health professionals who could be presumed to have some health background related to environmental health, however minimal". This cannot be explained by the purpose of the committee since all three executive orders of the different committees mentioned environmental public health related issues. Another finding of the authors is that a quarter of the opposed comments mentioned the possibility of bias in favor of gas industries in the composition of committees. The authors conclude saying that political leaders may not want to raise public health concerns not to handicap further economic development due to hydraulic fracturing.

Secondly, the conditions to allow hydraulic fracturing are being increasingly strengthened due to the move from governmental agencies' authority over the issue to elected officials' authority over it. The Shale Gas Drilling Safety Review Act of 2014 issued in Maryland[22] forbids the issuance of drilling permits until a high standard "risk assessment of public health and environmental hazards relating to hydraulic fracturing activities" is conducted for at least 18 months based on the Governor's executive order.

3.3   Institutional discourse and the public

A qualitative study using deliberative focus groups has been conducted in the North of England,[19] where there is a big shale gas reservoir exploited by hydraulic fracturing. These group discussions reflect many concerns on the issue of the use of unconventional oil and gas. There is a concern about trust linked with a doubt on the ability or will of public authorities to work for the greater social good since private interests and profits of industrial companies are seen as corruptive powers. Alienation is also a concern since the feeling of a game rigged against the public rises due to "decision making being made on your behalf without being given the possibility to voice an opinion". Exploitation also arises since economic rationality that is seen as favoring short-termism is accused of seducing policy-makers and industry. Risk is accentuated by what is hydraulic fracturing as well as what is at stake, and "blind spots" of current knowledge as well as risk assessment analysis are accused of increasing the potentiality of negative outcomes. Uncertainty and ignorance are seen as too important in the issue of hydraulic fracturing and decisions are therefore perceived as rushed, which is why participants favored some form of precautionary approach. There is a major fear on the possible disconnection between the public's and the authorities' visions of what is a good choice for the good reasons.

It also appears that media coverage and institutional responses are widely inaccurate to answer public concerns. Indeed, institutional responses to public concerns are mostly inadequate since they focus on risk assessment and giving information to the public that is considered anxious because ignorant. But public concerns are much wider and it appears that public knowledge on hydraulic fracturing is rather good.[19]

The hydraulic fracturing industry has lobbied for permissive regulation in Europe,[23] the US federal government, and US states.[24] On March 20, 2015 the rules for disclosing the chemicals used were tightened by the Obama administration.[25]

The new rules give companies involved 30 days from the beginning of an operation on federal land to disclose those chemicals.[25]

3.4 See also

- Hydraulic fracturing by country

3.5 References

[1] Nolon, John R.; Polidoro, Victoria (2012). "Hydrofracking: Disturbances Both Geological and Political: Who Decides?" (PDF). *The Urban Lawyer* **44** (3): 1–14. Retrieved 2012-12-21.

[2] Negro, Sorrell E. (February 2012). "Fracking Wars: Federal, State, and Local Conflicts over the Regulation of Natural Gas Activities" (PDF). *Zoning and Planning Law Report* (Thomson Reuters) **35** (2): 1–14. Retrieved 2014-05-01.

[3] Patel, Tara (31 March 2011). "The French Public Says No to 'Le Fracking'". *Bloomberg Businessweek*. Retrieved 22 February 2012.

[4] Patel, Tara (4 October 2011). "France to Keep Fracking Ban to Protect Environment, Sarkozy Says". *Bloomberg Businessweek*. Retrieved 22 February 2012.

[5] "LOI n° 2011-835 du 13 juillet 2011 visant à interdire l'exploration et l'exploitation des mines d'hydrocarbures liquides ou gazeux par fracturation hydraulique et à abroger les permis exclusifs de recherches comportant des projets ayant recours à cette technique"

[6] "Article L 110-1 du Code de l'Environnement"

[7] "Fracking ban upheld by French court". *BBC*. 11 October 2013. Retrieved 16 October 2013.

[8] Moore, Robbie. "Fracking, PR, and the Greening of Gas". *The International*. Retrieved 16 March 2013.

[9] Bakewell, Sally (13 December 2012). "U.K. Government Lifts Ban on Shale Gas Fracking". Bloomberg. Retrieved 26 March 2013.

[10] Hweshe, Francis (17 September 2012). "South Africa: International Groups Rally Against Fracking, TKAG Claims". *West Cape News*. Retrieved 11 February 2014.

[11] Nicola, Stefan; Andersen, Tino (26 February 2013). "Germany agrees on regulations to allow fracking for shale gas". Bloomberg. Retrieved 1 May 2014.

[12] "Commission recommendation on minimum principles for the exploration and production of hydrocarbons (such as shale gas) using high-volume hydraulic fracturing (2014/70/EU)". *Official Journal of the European Union*. 22 January 2014. Retrieved 13 March 2014.

[13] Healy, Dave (July 2012). Hydraulic Fracturing or 'Fracking': A Short Summary of Current Knowledge and Potential Environmental Impacts (PDF) (Report). Environmental Protection Agency. Retrieved 28 July 2013.

[14] Hass, Benjamin (14 August 2012). "Fracking Hazards Obscured in Failure to Disclose Wells". Bloomberg. Retrieved 27 March 2013.

[15] Soraghan, Mike (13 December 2013). "White House official backs FracFocus as preferred disclosure method". *E&E News*. Retrieved 27 March 2013.

[16] , Environmental Protection Agency

[17] Editorial Board (17 December 2014). "Gov. Cuomo Makes Sense on Fracking". *New York Times*. Retrieved 18 December 2014.

[18] de Rijke "Hydraulically fractured: unconventional gas and anthropology", *Anthropology today*, Volume 29, Number 2, April 2013

[19] Williams, Laurence, John "Framing fracking: public responses to potential unconventional fossil fuel exploitation in the North of England", Durham thesis, Durham University, 2014

[20] Goldstein, Kriesky, Pavliakova "Missing from the table: role of the environmental public health community in governmental advisory commission related to the Marcellus Shale Drilling", University of Pittsburgh, in *Environmental Health Perspectives*, Volume 120, Number 4, April 2012

[21] "Vermont Act 152"

[22] "Shale Gas Drilling Safety Review Act of 2014"

[23] Lipton, Eric; Hakim, Danny (October 18, 2013). "Lobbying Bonanza as Firms Try to Influence European Union". The New York Times.

[24] Kaplan, Thomas (November 25, 2011). "Millions Spent in Albany Fight to Drill for Gas". The New York Times.

[25] MATTHEW DALY; JOSH LEDERMAN (20 March 2015). "Politics Fracking: US Tightens Rules for Chemical Disclosure". ABC News Internet Ventures. Associated Press. Retrieved 20 March 2015.

3.6   External links

- The British Columbia (Canada) Oil and Gas Commission mandatory disclosure of hydraulic fracturing fluids

- Hydraulic Fracturing: Selected Legal Issues Congressional Research Service

Chapter 4

List of additives for hydraulic fracturing

In the United States, about 750 compounds have been listed as additives for hydraulic fracturing in a report to the US Congress in 2011 after originally being kept secret for "commercial reasons".[1][2] The following is a partial list of the chemical constituents in additives that are used or have been used in fracturing operations, as based on the report of the New York State Department of Environmental Conservation, some are known to be carcinogenic.[3]

In the UK only 'Non-Hazardous' chemicals are permitted for fracturing fluids by the Environment Agency. All chemicals have to be declared publicly and, increasingly, food additive based chemicals are available to allow fracking to take place safely.[4]

4.1   Additives used in the United States

4.2   See also

- Radionuclides associated with hydraulic fracturing

- Hydraulic fracturing proppants

- Hydraulic fracturing in the United States

- Hydraulic fracturing in the United Kingdom

4.3   References

[1] Nicholas Kusnetz (April 8, 2011). "Fracking Chemicals Cited in Congressional Report Stay Underground". ProPublica. Retrieved July 11, 2011.

[2] Chemicals Used in Hydraulic Fracturing (PDF) (Report). Committee on Energy and Commerce U.S. House of Representatives. April 18, 2011.

[3] "Natural Gas Development Activities and High-volume Hydraulic Fracturing" (PDF). New York State Department of Environmental Conservation. pp. 45–51.

[4]

Chapter 5

Baldwin Hills Dam disaster

This article is about the former reservoir in Baldwin Hills. For the lake in Arcadia, see Baldwin Lake (Los Angeles County, California).

The **Baldwin Hills Dam disaster** occurred on December 14, 1963, when the dam containing the **Baldwin Hills Reservoir** suffered a catastrophic failure and flooded the residential neighborhoods surrounding it. It began with signs of lining failure, followed by increasingly serious leakage through the dam at its east abutment. After three hours the dam breached, with a total release of 250 million US gallons (950,000 m^3), resulting in five deaths and the destruction of 277 homes. Vigorous rescue efforts averted a greater loss of life.

The reservoir was located on a low hilltop in Baldwin Hills, Los Angeles, California. It was constructed between 1947 and 1951 by the Los Angeles Department of Water and Power directly on an active fault line, which was subsidiary to the well known nearby Newport–Inglewood Fault. The underlying geologic strata were considered unstable for a reservoir, and the design called for a compacted soil lining meant to prevent seepage into the foundation. The fault lines were considered during planning but were deemed by some, although not all, of the engineers and geologists involved as not significant.[1]

The former reservoir is now part of the Kenneth Hahn State Recreation Area.

5.1   Significance and diagnoses of the failure

The failure of the Baldwin Hills Reservoir received an exceptional amount of attention from the civil engineering community and remains the subject of continuing interest. The reservoir had been conceived, designed, and built during and after World War II, a time when the pace of dam building was accelerating even as some disastrous dam failures were occurring, indicating a need for safer technologies. The builders of the Baldwin Hills dam, the Los Angeles Department of Water and Power, were aware of the difficult geologic conditions presented by the site and knew from past experiences, notably the catastrophic failure of the St. Francis Dam in 1928 in which nearly 600 people lost their lives, the serious consequences of a failure, even a small reservoir in an urban setting. But this was also an era of new engineering ventures on land, sea, and space, with new technologies boldly advanced to meet what were seen as hostile challenges from both nature and communist ideologies. While dams were recognized as potentially dangerous, like nuclear technologies, they were also considered by Americans as a showcase technology—a means of fending off danger and spreading progressive American technologies and associated social benefits at home and abroad.[2]

The Baldwin Hills dam designer, engineer Ralph Proctor, had also worked as an assistant civil engineer for the Los Angeles Department of Water and Power on the failed St. Francis Dam[3] and had subsequently devised new methods of producing compacted earth fill in building its replacement.[4] Proctor aggressively proceeded with the Baldwin Hills project even in the face of safety concerns and disagreements over important design details raised within his own department.[1]

Late 1963, when the failure occurred, was a time of another notable public disaster. Only two months before at the Vajont Dam in Italy, a massive landslide into the reservoir behind it, created a seiche which overtopped the dam flooding the valley below and causing the deaths of approximately 2000 people. The Baldwin Hills Reservoir had been built, as were others, to assure an ample supply of safe water for the people of Los Angeles in case of catastrophe such as

Baldwin Hills Reservoir after 1963 failure, view south. The gash through the dam corresponds to the alignment of a fault.

earthquake, fire, or war, and its failure was a blow to engineering confidence and the subject of many writings and two professional conferences (1972 and 1987, see references). The failure occurred shortly after the death of the authoritative Harvard engineer Karl Terzaghi whose ideas had long dominated both earth dam engineering and the engineering science of soil mechanics; Terzaghi had also made significant contributions to understanding subsidence in oilfields. This left the assessment of the Baldwin Hills failure in the hands of a new generation of engineers, some who took on conflicting roles as experts in various lawsuits.

The design and construction of the dam had been inspected and approved by the California Department of Water Resources. A meticulously documented study published by that agency in 1964—while pointing out various connections between oilfield operations in the Inglewood field and ground disturbances in the area, including beneath the reservoir and at some distance from the reservoir—concluded rather vaguely that the failure was due to "an unfortunate combination of physical factors".[5]

The monetary damages resulting from the failure were large, and some of the investigations which followed the state study were sponsored by litigants seeking more specific conclusions relevant to legal liability. This drew attention to oilfield operations in the area. From the outset it was clear that ground faulting and fault creep destroyed the reservoir, were probably related to the many feet of ground subsidence which had occurred a half mile west of the reservoir over decades of oil extraction in the Inglewood field. The oilfield-related subsidence in the Inglewood field, though generally denied by the oil companies as a legal policy, was documented exhaustively by the US Geological Survey in 1969.[6] Subsidence following oil extraction from shallow deposits in unconsolidated sediments had been understood by oil industry experts since the 1920s.[7]

Following the discovery in 1970 by geologist Douglas Hamilton of faulting and surface seepage of oilfield waste brines along the fault which traversed and extended south of the reservoir, Hamilton and Meehan concluded that oilfield injection for waste disposal and improved recovery of oil, a new technology at the time, was a significant cause of the failure, triggering hydraulic fracturing and aggravating movements on a fault traversing the reservoir even on the day of the failure.[8] Subsequently the US Geological Survey concluded in 1976 that displacements at the ground surface causing reservoir failure and also ground cracking in the Stocker-LaBrea area southeast of the reservoir were 90 percent or more attributable to exploitation of the Inglewood oil field, and that this faulting was likely aggravated by waterflooding with pressures exceeding hydraulic fracturing levels.[9]

By 1972, nearly a decade after the failure, the immediate legal issues had been settled out of court and the matter was reopened as a topic of discussion among investigators in a published engineering conference at Purdue University.

Engineer Thomas Leps, who had served as consultant on the 1964 state investigation, took on a role as neutral reviewer in this and most subsequent American studies of the failure. Leps concluded that there had been about 7 inches of offset on the fault beneath the reservoir during its life, about 2 inches of which had occurred in the months just before the failure. Leps associated the latter with repressurization of the oilfield. This, along with stretching of the ground due to subsidence of about 12 feet from oil extraction, had caused the lining failure which doomed the reservoir.[10]

Some prominent consultants including those on a team led by Arthur Casagrande, Harvard successor to Karl Terzaghi, held that oilfield operations were not a significant influence at all but that the failure was the result of defective siting and design with the heavy weight of the dam and reservoir being the significant cause of the fatal foundation movement.[11] This view exonerated the oil companies, namely Standard Oil, which had sponsored the study. Casagrande refused to acknowledge any ground movements in the area as being related to oilfield operations and argued that ground movements that affected the dam were found only beneath the reservoir, not in adjoining areas.

Most of these questions were examined once again in 1986 following investigations of a suspiciously similar major failure of the Bureau of Reclamation's Teton Dam in June, 1976, and a near failure of the Department of Water and Power's Lower Van Norman Dam in the 1971 San Fernando earthquake. Professor Ronald Scott of Caltech, who had participated in the Casagrande studies, noted at a follow-up 1987 conference on Baldwin Hills[1] that Casagrande had ignored or been unaware of ground movements clearly unrelated to the reservoir (e.g. those at Stocker-LaBrea) in his analysis. Another engineer, Stanley Wilson—who had also worked with Casagrande on the 1972 studies and supported the claim that oilfield subsidence was an insignificant cause—now conceded that analogous ground offsets extended well outside the reservoir area, notably in the Stocker-LaBrea area, so that the reservoir and other fault movements could not be attributed to the reservoir itself—thus tacitly attributing responsibility for the failure to oilfield operations. Hence, there appeared to be convergence of opinion on the role of oilfield subsidence and repressurization.

The issue of oilfield causation was a central theme in most of these discussions, with little attention having been directed to the details of the failure. The absolute necessity of a lining for this site was generally taken for granted in these proceedings even as it had been by Proctor himself, regardless of the fact that almost all earth dams perform satisfactorily without linings. Some suggestions as to possible preventive design and construction techniques that might have made the dam safer were raised as engineering consensus reached a state of textbook knowledge in the late 1980s.[12] For example, the character of the compacted earth lining (which had been regularly referred to as clay but must have been substantially silt and sand, having been derived from the local Inglewood formation[5]) was raised, if obliquely, in the suggestion made in the end that improved performance might have come from the use of a different lining material.[13]

In 2001 a new angle on failure analysis was introduced by Mahunthan and Schofield, who concluded that overcompaction of the dam fill and lining was a significant aggravating factor in both the Baldwin Hills and Teton failures.[14] This assertion was based on Schofield's concepts of critical-state soil mechanics,[15] a corollary of which was that heavily compacted but lightly confined soils could be dangerously unstable where seepage forces were present. This issue had not been raised in the previous American-dominated discussions and remains in some degree contrary to American ideas in both theoretical soil mechanics and practical geotechnical engineering. In fact the 1964 DWR failure study implied that heavy compaction was a favored technique for earth dam construction,[16] and this assumption appeared not to have been reexamined over the twenty five years of post-failure investigation and discussion.

The failure of the reservoir has been a subject of ongoing interest in the field of dam breach studies. A recent study examined the dam failure as a two-stage process and succeeded in modeling the flood in the urban area downstream.[17]

Although the Baldwin Hills Reservoir site has now been dedicated as a community park, and there is no further significant hazard associated with ground movements there, the associated faults to the southeast (Stocker-LaBrea and the Windsor School area) continue to move significantly as of 2012, causing damage to private and public facilities. The current oilfield operator, Plains Exploration and Production Company (PXP), which has intensified production and development efforts in the oilfield with the rising price of petroleum, does not, unlike its predecessor Standard Oil, acknowledge any causal connection between fault movements and oilfield activities, and has retained a team of consultants who support this position or conclude that the causes of the movements are unknown.[18] The role of shallow hydraulic fracturing, which has recently been introduced as a means of stimulating production at depths of about 2000 feet in the southeast part of the Inglewood field,[19] and at greater depths elsewhere in the field, has also generated public concern and controversy. However, oil operators, while admitting that fracture pressures[20][21] are being exceeded,[19] do not acknowledge a relationship between injection at fracturing pressure levels and fault movement. The PXP and PXP consultant conclusions, that adverse effects are either unknown or not present, are disputed by other reviewers.[22]

Recent discharges of oilfield gases in the Baldwin Hills may also be related to raised pressures resulting from injection, and may be of similar origin as the gas problems in the nearby Salt lake field.[23]

5.2 Coverage

KTLA used a helicopter to cover the disaster. Common today, this was perhaps the first such live aerial coverage of a breaking news event.[24]

5.3 See also

- List of lakes in California

5.4 Notes

[1] Scott 1987

[2] Meehan, RL 2011

[3] Coroner's Inquest 1928

[4] Rogers 2011

[5] California 1964

[6] Castle 1969

[7] Geertsma 1973

[8] Hamilton 1971

[9] Castle and Yerkes 1976

[10] Leps 1972 p541

[11] Casagrande 1972

[12] James Ed Al 1988

[13] James et al 1988

[14] Muhunthan and Schofield 2001

[15] Schofield 2006

[16] California 1964 p 11 and Table V-2

[17] Gallegos et al 2009

[18] StrataGen Engineering 2012

[19] Moodie 2004

[20] Hubbert 1957

[21] Castle 1976

[22] Meehan 2012

[23] Hamilton 1992

[24] Pool, Bob (December 11, 2003). "Serene Hilltop Marks Site of Landmark Disaster". *Los Angeles Times*.

5.5 References

- California Department of Water Resources (April 1964). "Investigation of Failure Baldwin Hills Reservoir".

- Casagrande, A,; Wilson,, SD,; Schwantes, ED, (1972). "The Baldwin Hills Reservoir failure in retrospect". *Proceedings of the ASCE Specialty Conference on the Performance of Earth and Earth-Supported Structures, Purdue University*.

- Castle, RO; Yerkes, RF (1969). "A Study of Surface Deformations Associated with Oil-Field Operations:". *Report of the U.S. Geological Survey, Menlo Park, California*.

- Castle, RO; Yerkes, RF (1976). "Recent Surface Movements in the Baldwin Hills, Los Angeles County California". *Geological Survey professional paper 882*.

- Cowen, Richard (11 February 2002). "CHAPTER XX: Man-made Subsidence". University of California, Davis. Retrieved 2009-04-05.

- Fireman's Grapevine (February 1964). "Firemen Save 18 Lives in Baldwin Hills Flood". *The Fireman's Grapevine*. Retrieved 2009-04-05.

- Gallegos, HA,; Sanders,, BF,; Schubert, JE, (August 2009). "Two-dimensional, high-resolution modeling of urban dam-break flooding: A case study of Baldwin Hills, California". *Advances in Water Resources Vol 32, Issue 8*.

- Geertsma, J (1973). "Land subsidence above compacting oil and gas reservoirs". *Journal of Petroleum Technology SPE_AIME*.

- Hamilton, DH; Meehan, RL (23 April 1971). "Ground Rupture in the Baldwin Hills" (PDF). *Science* **172** (3981): 333–344. doi:10.1126/science.172.3981.333. PMID 17756033. Retrieved 2009-04-05.

- Hamilton, DH; Meehan, RL (1992). *"Cause of the 1985 Ross Store Explosion and Other Gas Ventings, Fairfax District, Los Angeles,"* Engineering geology practice in southern California; ed. by Bernard W. Pipkin and Richard J. Proctor. Association of Engineering Geologists. Southern California Section; also presented June 19–22, 2000, AAPG Pacific Section and Western Region Society of Petroleum Engineers Meeting in Long Beach, California. p. 769. ISBN 0-89863-171-8.

- Hubbert, Wk and Willis, DG (1957). "Mechanics of Hydraulic Fracturing". *AIM Peroleum Transactions* (TP 5497).

- James, L. B.; Kiersch, G. A.; Jansen, R. B.; Leps, T. M. (1988). ""Lessons from notable events."". *in Jansen, RB ed. "Advanced dam engineering for design, construction, and rehabilitation"*. NY.: Van Nostrand Reinhold.

- Leps, Thomas M, (1972). "Analysis of failure of Baldwin Hills Reservoir". *Proceedings of the ASCE Specialty Conference on the Performance of Earth and Earth-Supported Structures, Purdue University*.

- "The Water Damage restoration plan". UAC water restoration Los Angeles division. 2008-05-12.

- Meehan, RL (2012). "Ground rupture in the Baldwin Hills: fracking 2012". Retrieved 2012-12-15.

- Moodie, WH; et al. (2004). "Multistage Oil-Base Frac-Packing in the Thick Inglewood Field". *Society Peroleum Engineers* (SPE 90975): 9p.

- Muhunthan; Schoefield. "Liquefaction and Dam Failures, GeoDenver-2000, Denver, Colorado".

- Rogers, David (2011). "Mechanical Compaction of Soils for Engineering Purposes" (PDF). Retrieved 2011-06-15.

- Schofield, A. N. (2006). *Disturbed soil properties and geotechnical design*. Thomas Telford. p. 216. ISBN 978-0-7277-2982-8

- Scott, RF (1987). "Baldwin Hills reservoir failure in review". *Engineering Geology* **24** (1–4). ISSN 0013-7952.

- Stratagen Engineering Company (2012). *PXP Baldwin Hills Inglewood Oilfield: Review and Discussion of 2012 Surface Survey Results*. Consultant Report to PXP.

- Zhai, Zongyu; Sharma, Mukul (2005). "A new approach to modelling hydraulic fracturing in unconsolidated sands". *Society of Petroleum Engineers*. SP96246: 14.

- Coroner's Inquest (1928). *Transcript of Testimony and Verdict of the Coroner's Jury In the Inquest Over Victims of St. Francis Dam Disaster: Book 26902*. Los Angeles County Department of Coroner. p. 139.

5.6 External links

- The History Channel segment about the disaster
- Ground Rupture in the Baldwin Hills
- "Mechanical Compaction of Soils for Engineering Purposes"
- Study & Task Force Report/API paper.pdf"Ross Store Explosion"

Coordinates: 34°00′29″N 118°21′49″W / 34.008158°N 118.363577°W

Chapter 6

Canadian Association of Petroleum Producers

The **Canadian Association of Petroleum Producers** (CAPP), with its head office in Calgary, Alberta, is an influential lobby group that represents the upstream Canadian oil and natural gas industry.[1] CAPP's members produce "90% of Canada's natural gas and crude oil"[2] and "are an important part of a national industry with revenues of about $100 billion-a-year (CAPP 2011)."[2]

6.1   History

CAPP origins can be traced back to the Alberta Oil Operators' Association, which was founded in 1927, after the discovery of the Turner Valley Oil Field. In 1947, the Alberta Petroleum Association changed its name to the Western Canadian Petroleum Association, and In 1952, the Western Canada Petroleum Association amalgamated with the Saskatchewan Operators' Association and adopted the name Canadian Petroleum Association.

At a meeting on December 9, 1952, the CPA drafted a new constitution which outlined the objectives of the organization as follows:

- to establish better understanding between the petroleum and natural gas industry and the public

- to encourage cooperation between the petroleum and natural gas industry and federal, provincial and local governments, and other authoritative bodies

- to provide a forum for the discussion of matters affecting the welfare of its members

- to foster better understanding between the Association and purposes

On June 10, 1958 the CPA opened an office in Ottawa and became "one (of) the oldest, largest and most influential lobby groups in Canada."[3] It provided the federal government with information pertaining to the oil industry while keeping the CPA informed about political trends, government regulations and statistics. By 1965 the CPA had a membership of more than 200 members representing roughly 97 percent of all oil and gas production in Canada. In 1981, two years after the first commercial discovery at Hibernia off the coast of Newfoundland, the CPA opened an office in St. John's in cooperation with the Eastcoast Petroleum Operators' Association.

In 1992, when the Canadian Association of Petroleum Producers (CAPP) was formed, with the CPA amalgamation with the Independent Petroleum Association of Canada (IPAC) to form the Canadian Association of Petroleum Producers (CAPP),[3] Gerry Protti was named as founding president.[4][notes 1]

6.1.1   CAPP Hall of Fame

- Michael (Jeffery) A. Jackson (1914–1972), founding director of the Independent Petroleum Association of Canada, lobbied for the western Canadian oil industry and pushed for pipelines to Central Canada.[5]

6.2   Advocacy for oil industry

Canada's estimated total oil reserves including conventional oil were approximately 180 billion barrels (29 km^3), behind only Saudi Arabia and Venezuela. Canada produces approximately 2.7 million barrels (430,000 m^3) of crude oil a day, and 6.4 trillion cubic feet (180 km^3) of natural gas per year. In 2013, an IPSOS poll showed a majority (75%) of Canadians prioritize local crude before using imported oil from foreign sources, while just over one in ten (14%) 'disagree' (4% strongly/11% somewhat) and 11% have no opinion.[6]

CAPP has advocated for the industry as GHG emissions rose 14% in 2009 and 2010, by its own admission. . However, GHG emissions per barrel of oil sands crude produced have dropped by 26% since 1990 as a result of new operating practices and technology.

According to IHS CERA, oil sands crude has similar CO_2 emissions to other heavy oils and is 9% more intensive than the U.S. crude supply average on a wells-to-wheels basis.[7]

The industry employs 550,000 people and paid billions in taxes and royalties to different levels of government.

Advocacy for Oil Sands CAPP's series of meetings in 2010 in eight cities in Canada and the United States, including Vancouver, Edmonton, Ottawa, Toronto, Montreal, Washington D.C., New York and Chicago, with CAPP representatives, oil sands CEOs and 160 key stakeholders, culminated in a report entitled *Dialogues* published on 14 April 2011.[2]

6.2.1   Advocacy for fracking

CAPP advocates for the use the controversial technology hydraulic fracturing. In 2010 released a series of voluntary Guiding Principles for Hydraulic Fracturing for Canadian natural gas producers to adhere to. The Guiding Principles of Hydraulic Fracturing were followed in 2011 by an agreed set of Six Hydraulic Fracturing Practices for: 1. Fracturing fluid additive disclosure 2. Fracturing fluid additive risk management 3. Baseline groundwater testing 4. Wellbore construction 5. Water sourcing and reuse 6. Fluid handling, transport, disposal

[8]

6.2.2   Criticisms and concerns

The Council of Canadians and Sierra Club Canada take a strong position against hydraulic fracturing[9] and want it banned in Canada entirely, and have supported specific bans in Nova Scotia [10] and New Brunswick.

6.3   Advocacy for Crude Oil Exports via Canada's West Coast

CAPP supports and advocates for exports of Canadian crude oil via Canada's west coast via the Northern Gateway and the KinderMorgan TransMountain Expansion Project. In September 2011, the Asia Pacific Foundation of Canada (APF Canada) and the Canada West Foundation established the Canada-Asia Energy Futures Task Force with Kathleen (Kathy) E. Sendall, C.M., FCAE,[notes 2] a former Governor and Board Chair of the Canadian Association of Petroleum Producers (CAPP) and Kevin G. Lynch, a Canadian economist and former Clerk of the Privy Council and Secretary to the Cabinet, Canada's most senior civil servant as co-chairs, to investigate a long-term Canada-Asia energy relationship. One of their recommendations was the creation of a public energy transportation corridor.[11]

6.3.1   Criticisms and concerns

Canadian opponents to the Northern Gateway , intended to permit shipping of high-carbon Canadian crude over ecologically sensitive rivers and waters to carbon-uncontrolled countries including India and China, include 61 First Nations in British Columbia.

6.4   Advocacy for Keystone XL Pipelines expansion

CAPP supports and advocates for the $7-billion pipeline expansion project by the Canadian-based company TransCanada to build the Keystone XL, that would extend and expand capacity of existing pipelines, that transport crude oil from the Athabasca oil sands in northern Alberta to tidewater and to refineries in the Gulf, capable of refining the heavy bitumen crude oil.

6.4.1   Criticisms and concerns

Nine winners of the Nobel Peace Prize, including Archbishop Desmond Tutu and the Dalai Lama, were signatories to a letter to pressure U.S. President Barack Obama to reject the $7-billion pipeline expansion project by the Canadian-based company TransCanada to build the Keystone XL.

The position of the Nobel Peace Prize winners, essentially, is that one rich nation selling increasingly heavy high-carbon oil to another sabotages any effort to reach a deal on global carbon controls, and that moves to expand this export (like Keystone XL or Northern Gateway) cause significant and direct risks to world peace, as climate victim countries become subject to chaotic weather, fighting over scarce water (especially in Southeast Asia and Africa), flooding and rising sea levels.

6.5   Advocacy regarding GHG emissions

CAPP opposed the Kyoto Protocol, from which Stephen Harper withdrew Canada in December 2011. CAPP's lobbying efforts included favouring "made in Canada" approach and advocating for a carbon-pricing program. In 2007 a carbon tax was implemented in Alberta, Canada's major oil and gas producing province. Supported by CAPP and in the industry, the $15/tonne carbon tax feeds a GHG emissions reduction technology fund.

By 2008, the oil sands industry contributes (approximately 3%–4%) of Canada's GHG emissions (approximately 3%–4%. By 2012, oil sands contributed 0.14% of global GHG emissions. Transportation and electricity were the largest contributors of GHG, with transportation contributing 190 Mt of CO_2 equivalent per year ($MtCO_2eq$ yr−1) and electricity and heat generation: 125 $MtCO_2eq$ yr−1. However, by 2007 (Environment Canada 2007) cautioned that unrestricted development of the oil sands could increase its emissions and the percentage. A 2008 CAPP report argued that both the Alberta and Federal governments adopted "comparable industry GHG emissions targets in which large emitters must reduce their emissions by either improving their operation, purchasing emissions credits or investing in technology funds."[1][12]

6.5.1   Criticisms and concerns

Canada was the first signatory nation to walk away from the Kyoto Protocol in 2012. The U.S. abandoned the Kyoto Protocol in 2001.[13]

6.6 CAPP initiative 2011: Oil and gas, industry, provincial regulators collaborate on strategies to shape public perception of fracking, water use and shale gas development

In the summer of 2011 CAPP contacted ENV to requested a meeting with the Canadian Society for Unconventional Gas (CSUG), and officials from several government ministries, including Alberta Environment, Energy, Sustainable Resource Development (SRD), as well as the Energy Resources Conservation Board (ERCB), (now Alberta Energy Regulator) to discuss CAPP's desire to strike a committee to develop a public communications strategy focused on fracturing and water use associated with shale gas development."[14] Senior-level government and industry officials attended the joint meeting "to develop a plan to shape public perceptions of shale gas development and water use." From Alberta Energy participants included Director of Unconventional Gas Doug Bowes, Associate Branch Head Matthew Foss, Environment and Resource Services Audrey Murray, Executive Director of Resource Development Sharla Rauschning, Assistant Deputy Minister Resource Development Policy Division Jennifer Steber. From Alberta Environment participants included, Deputy Minister Ernie Hui, Former Head of Groundwater Policy within the Water Policy Branch, now the Exec. Dir. of OH&S Policy and Program with Human Services Ross Nairne. From Sustainable Resource Development (SRD) participants included Assistant Deputy Minister Glen Selland, Executive Director, Land Management Branch Jeff Reynolds, Officials from CAPP included VP Operations David Pryce, Manager of BC Operations Brad Herald, Manager of Water and Reclamation Tara Payment. From the Canadian Society for Unconventional Gas (CSUG) CSUG (a.k.a. CSUR) participants included Vice President Kevin Heffernan.

June 8, 2011, e-mail to senior government officials from the Energy Resources Conservation Board, the arm's length regulator of the oil and gas industry in Alberta, to several meetings to produce a collaborative communications campaign on fracking strategy. On 9 June 2011 the Alberta government approved collaborative communications campaign in the minutes of their joint meeting.[15] stating that

(Government of Alberta) agrees communication is a priority including a joint industry/GOA committee to develop similar language and terminology for discussion of shale gas issues and operations... The objective is to improve public understanding of shale gas operations and improve public knowledge and confidence. Preparation of a common background information document may be of value (when) targeted at a public audience.

Government of Alberta, Joint meeting with CAPP held 9 June 2011

6.6.1 Criticisms and concerns

By 29 November 2011, the CBC and the Alberta Federation of Labour (AFL), were investigating the role played by CAPP in influencing Alberta Environment over public communications surrounding shale gas extraction, a controversial practice that has significant environmental concerns associated with it, especially when fracturing is employed. Questions were raised about the legality of private interests influencing government. Complaints were filed and dismissed.[16]

6.7 Selected CAPP publications

- [12]

- CAPP Report: Gas Use by the Canadian Oil Sands Industry (Report). Natural Gas. Canadian Association of Petroleum Producers (CAPP). December 2007. p. 22.

- CAPP 2008 Facing our Challenges-2007 Stewardship Report (Report). Canadian Association of Petroleum Producers. 2008.

- Overview: Water Use in Canada's Oil Sands (Report). Canadian Association of Petroleum Producers (CAPP). September 2010.p. 1.

- GHG 101 (Report). Canadian Association of Petroleum Producers (CAPP). 2009.

- Crude Oil Forecast, Markets & Pipeline Expansions 101 (Report). Canadian Association of Petroleum Producers (CAPP). June 2010.p. 30

- Overview: Greenhouse Gas Emission in Canada's Oil Sands (Report). Canadian Association of Petroleum Producers (CAPP). August 2010.p. 1

- Report of the Dialogues on the Oil Sands: Engaging Canadians and Americans (Report). Calgary, Alberta: Canadian Association of Petroleum Producers (CAPP). 14 April 2011. Full text on-line report

6.8 See also

- American Petroleum Institute (API)

6.9 Notes

[1] First Nations, landowners, ranchers and Alberta Surface Rights among others oppose the appointment of Gerry Protti as chair of the newly created regulatory group Alberta Energy Regulator which takes over responsibilities of the Energy Resources Conservation Board (ERCB) and other key institutions in terms of regulatory issues.

[2] Kathleen Sendall, is director CGG (Paris, France), Director of Enmax Corporation (Calgary, AB); Vice Chair, Alberta Innovates – Energy and Environment Solutions (Calgary, AB); Co-Chair, Canada West/Asia Pacific Foundation Task Force (Calgary, AB)." In 2013 Prime Minister Harper appointed her the Sustainable Development Advisory Council (SDAC) and the Advisory Council for Promoting Women on Boards. She advises "federal and provincial governments in the areas of climate change, carbon capture and storage, environmental legislation, and Arctic foreign policy, and recently chaired the Canadian Council of Academies Assessment Panel on the State of Industrial R&D in Canada. Previously, Ms. Sendall led Petro-Canada's North American Natural Gas Business Unit."Kathleen Sendall Biography

6.10 References

[1] Alex D Charpentier; Joule A Bergerson; Heather L MacLean (2009). "Understanding the Canadian oil sands industry's greenhouse gas emissions" (PDF). *Environmental Research Letters* (IOP Publishing Ltd). doi:10.1088/1748-9326/4/1/014005.

[2] Report of the Dialogues on the Oil Sands: Engaging Canadians and Americans (Report). Calgary, Alberta: Canadian Association of Petroleum Producers (CAPP). 14 April 2011.

[3] "History of CAPP". Retrieved 30 May 2012.

[4] Weber, Bob (5 March 2013). "Gerry Protti, New Alberta Energy Regulator Head, Not The Right Man For The Job: Critics". Canadian Press. Retrieved 25 June 2013.

[5] "CAPP Hall of Fame". Canadian Association of Petroleum Producers.

[6] IPSOS North America (October 13, 2013). "Three-Quarters (75%) of Canadians Believe Oil Refineries Should Prioritize Local Oil Before Importing From Other Countries". *IPSOS*.

[7] Oilsands Review. "Study Says 45% Of U.S. Oil Supply Of Similar GHG Intensity As Oilsands". *www.oilsandsreview.com*. Glacier Media.

[8] http://landusekn.ca/resource/guiding-principles-hydraulic-fracturing-capp

[9] http://www.canadians.org/water/issues/fracking/index.html

[10] http://www.sierraclub.ca/en/node/4670

[11] Securing Canada's energy future: report of the Canada-Asia energy futures task force (PDF) (Report). Asia Pacific Foundation of Canada. June 2013.

[12] CAPP 2008 Facing our Challenges-2007 Stewardship Report (Report). Canadian Association of Petroleum Producers. 2008.

[13] Greenpeace. http://www.greenpeace.org/usa/en/news-and-blogs/news/u-s-withdraws-from-kyoto-prot/. Missing or empty |title= (help)

[14] "email sent by Doug Bowes, Director of Unconventional Gas, Department of Energy on 8 June 2011 in an e-mail to senior government officials". Documents On The CAPP/ERCB/SRD Fracking Relationship. Alberta Surface Rights. Retrieved 25 June 2013.

[15] Rusnell, Charles (29 November 2011). "Alberta worked with industry on fracking PR strategy". CBC News.

[16] "Illegal lobbying complaint against CAPP dismissed". *CBC News*. 28 March 2012. Retrieved 20 August 2012.

6.11 Further reading

Additional information about the lobbying controversy can be found here: http://www.cbc.ca/news/canada/edmonton/story/2011/11/29/edmonton-lobbying-compalint-dismissed.html

- Mech, Michelle (May 2011). A Comprehensive Guide to the Alberta Oil Sands: Understanding the Environmental and Human Impacts, Export Implications, and Political, Economic, and Industry Implications, and Political, Economic, and Industry Influences (PDF) (Report). Green Party.

6.12 External links

- CAPP web site

Chapter 7

Canol shale play

The **Canol shale play** is the name for a region of Canada's Northwest Territories that is being investigated as a potential source of shale oil.[1][2] The region centers around the known reserves of conventionally exploitable petroleum at Norman Wells.

Oil industry commentators see the region as a potentially rich source of future revenue. Local native groups are concerned over the regions remoteness and extreme environmental sensitivity, and to the lack of any testing standards to scientifically measure whether the use of toxic chemicals in hydraulic fracking is causing irredeemable damage to groundwater overlaying the regions of fractured rock.[3][4] The *Northern Journal quoted some of the objections of Sahtu elder Madeline Karkagie:*

The first test fracturing was conducted in the region in 2012, without any prior environmental assessments being performed.[4] On October 6, 2013, the *Globe and Mail* reported that Henry Sykes, the CEO of MGM Energy, complained that environmental concerns were causing financially troubling delays.[5] Canada's National Energy Board gave Conoco Phillips permission to build fracking wells in the region on October 30, 2013.[6] On January 17, 2014, the *Globe and Mail* reported drilling would begin in February.[7]

The Canol pipeline (short for "Canadian Oil") was a small diameter pipeline completed by the US Army in 1944 to help supply conventional crude oil from Norman Wells to Alaska. It was built as a short term project during World War II, and was operated barely more than a year.

7.1   References

[1] Darren Campbell (2013-05). "Could the N.W.T.'s Canol shale be the next Bakken?". Alberta Oil magazine. Archived from the original on 2013-06-11. Retrieved 2014-08-18. These days Hogg should be having a lot of fun. MGM Energy – a small Calgary-based junior – is part of a group of companies leading the charge to open up the N.W.T's Canol shale oil play. No one is sure just how much oil might be trapped in the dense rock that lines the remote central Mackenzie Valley where the play is located, but its potential conjures up images of North Dakota's Bakken, which was producing over 700,000 barrels of oil per day in 2012. Check date values in: |date= (help)

[2] Jack Danylchuk (2013-09-30). "Canol play sees new competitor on the Mackenzie by". Northern Journal. Archived from the original on 2013-10-29. Retrieved 2014-08-18. "The Canol Shale play was one of many potential business opportunities which was part of our decision making," Gram said, but emphasized the company's "general strategy is based more on trying to satisfy local markets. If some of these bigger plays came to fruition, we will be in position if there was a fit for ITB."

[3] Jack Danylchuk (2014-01-06). "Sahtu fracking opponents calling for a region-wide vote". Northern Journal. Archived from the original on 2014-08-18. Opponents to hydraulic fracturing in the Sahtu want residents to have the last word on the controversial process and the future of the Canol shale play.

[4] Derek Leahy (2014-02-27). "NWT Residents Demand Environmental Reviews Before Fracking Is Permitted". Desmog. Retrieved 2014-02. Residents of the Northwest Territories are demanding environmental reviews be conducted before companies are permitted to 'frack' for oil in the NWT. Despite controversy in Canada and other countries around the effects fracking or hydraulic fracturing has on water and climate change, the NWT's first fracking project was approved last October without an environmental assessment. Check date values in: |accessdate= (help)

[5] Jeffrey Jones (2013-10-06). "NWT's shale hopes dimmed by slow approvals, explorer says". Globe and Mail. Archived from the original on 2014-08-12. Retrieved 2014-08-18. MGM Energy Corp. chief executive officer Henry Sykes said regulatory hurdles are the main reason he believes that production growth in the Northwest Territories' Canol shale play in the Central Mackenzie Valley will not mirror the industry's success in the prolific Bakken oil fields of North Dakota.

[6] Carrie Tait (2013-01-30). "ConocoPhillips wins first regulatory approval of fracking in the North". Globe and Mail. Archived from the original on 2014-08-12. Retrieved 2014-08-18. The National Energy Board gave ConocoPhillips Co. the right to drill two wells and then use horizontal fracturing, known as fracking, methods to extract oil out of shale rocks. This is the first time the NEB has authorized a company to frack in the North.

[7] Jeffrey Jones, Carrie Tait (2014-01-17). "Fracking and climate change: Canada's Far North gets an energy boost". Globe and Mail. Archived from the original on 2014-08-12. Retrieved 2014-08-18. The global oil and gas company is preparing to employ horizontal drilling and hydraulic fracturing technology on wells in the Canol shale oil play in the Central Mackenzie Valley in February.

Chapter 8

Chevron CRUSH

Chevron CRUSH is an experimental *in situ* shale oil extraction technology to convert kerogen in oil shale to shale oil. The name stands for Chevron's Technology for the Recovery and Upgrading of Oil from Shale. It is developed jointly by Chevron Corporation and the Los Alamos National Laboratory.[1]

8.1 History

The Chevron CRUSH technology bases on the earlier *in situ* efforts. Sinclair Oil Corporation conducted an experiment using both natural and induced fractures to establish communication between wells and developing an *in situ* combustion process.[2][3] Geokinetics, the Sandia National Laboratories, and the Laramie Energy Technology Center of the U.S. Department of Energy conducted field tests fracturing oil-shale formation by explosives and hydraulic fracturing technology.[2] Equity Oil Company, Continental Oil Company and the University of Akron studied the benefit of carbon dioxide as a carrier gas to facilitate a higher yield of shale oil.[2] Based on these works, Chevron Corporation and the Los Alamos National Laboratory started a cooperation in 2006 to improve the recovery of hydrocarbons from oil shale.[1] In 2006, the United States Department of the Interior issued a research, development and demonstration lease for Chevron's demonstration project on public lands in Colorado's Piceance Basin.[4] In February 2012, Chevron notified the Bureau of Land Management and the Department of Reclamation, Mining and Safety that it intends to divest this lease.[5][6]

8.2 Process

For decomposition kerogen in oil shale, the Chevron CRUSH process uses heated carbon dioxide. The process involves drilling vertical wells into the oil shale formation and applying horizontal fractures induced by injecting carbon dioxide through drilled wells and then pressured through the formation for circulation through the fractured intervals to rubblize the production zone. For further rubblization propellants and explosives may be used. The used carbon dioxide then be routed to the gas generator to be reheated and recycled.[2][7][8] The remaining organic matter in previously heated and depleted zones is combusted *in-situ* to generate the heated gases required to process successive intervals. These gases would then be pressured from the depleted zone into the newly fractured portion of the formation and the process would be repeated.[2] The hydrocarbon fluids are brought up in conventional vertical oil wells.[2][7][8]

8.3 Isolation of groundwater

The processing area is isolated from surrounding groundwater by creating fractured areas ("pockets"), approximately 1 to 5 acres (4,000 to 20,200 m^2) wide and 50 feet (15 m) high within the center of the oil shale deposit. In this way, about 75 feet (23 m) of the confining layer would separate the process area from the water bearing layers above and below, keeping the aquifers out of the production zone.[2]

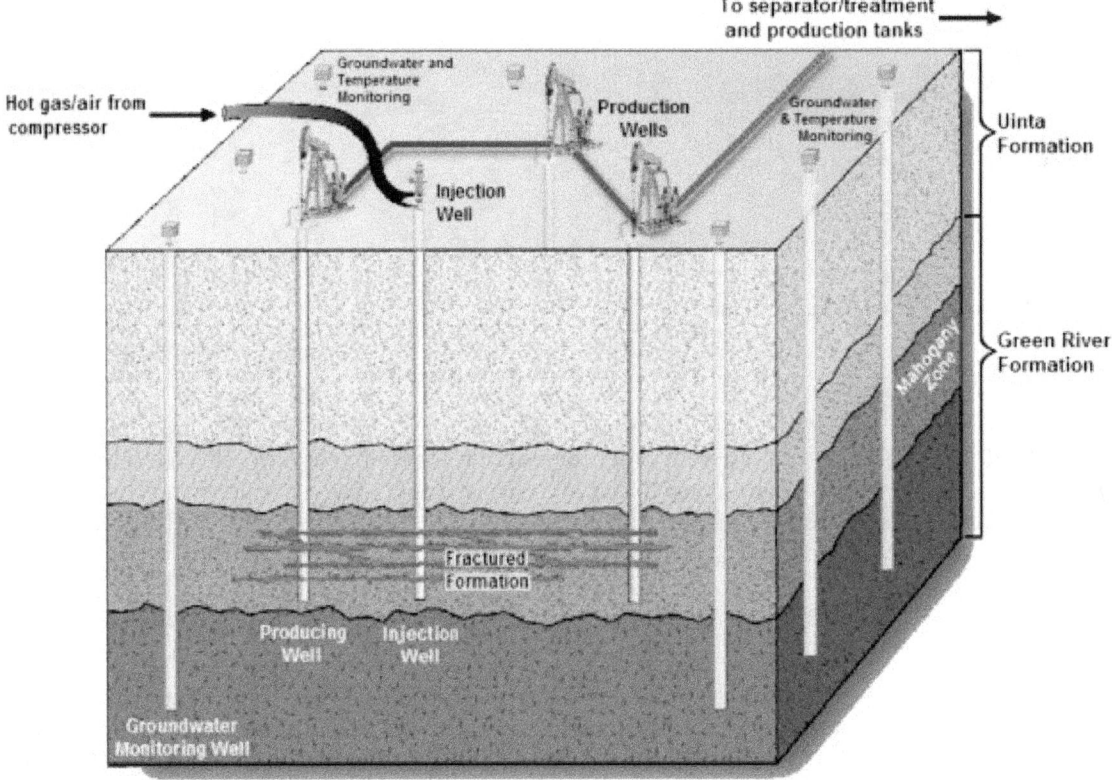

Chevron CRUSH process

8.4 See also

- Shell in situ conversion process
- ExxonMobil Electrofrac

8.5 References

[1] "Chevron and Los Alamos Jointly Research Oil Shale Hydrocarbon Recovery". Green Car Congress. 2006-09-25. Retrieved 2009-04-12.

[2] "Environmental assessment. Chevron Oil Shale Research, Development & Demonstration" (PDF). U.S. Department of Interior. Bureau of Land Management. White River Field Office. November 2006. CO-110-2006-120-EA. Retrieved 2009-04-12.

[3] Lee, Sunggyu (1991). *Oil Shale Technology*. CRC Press. p. 124. ISBN 0-8493-4615-0.

[4] "Interior Department Issues Oil Shale Research, Development and Demonstration Leases for Public Lands in Colorado" (Press release). Bureau of Land Management. 2006-12-15. Retrieved 2009-04-12.

[5] "Chevron leaving Western Slope oil shale project". *Denver Business Journal*. 2012-02-28. Retrieved 2012-03-12.

[6] Hooper, Troy (2012-02-29). "Chevron giving up oil shale research in western Colorado to pursue other projects". *The Colorado Independent*. Retrieved 2014-03-31.

[7] "Secure Fuels from Domestic Resources: The Continuing Evolution of America's Oil Shale and Tar Sands Industries" (PDF). United States Department of Energy. 2007. pp. 1–68. Retrieved 2007-07-11.

[8] "Oil Shale Research, Development & Demonstration Project. Plan of Operation" (PDF). Chevron USA Inc. 2006-02-15. Retrieved 2009-04-12.

Chapter 9

Environmental impact of hydraulic fracturing

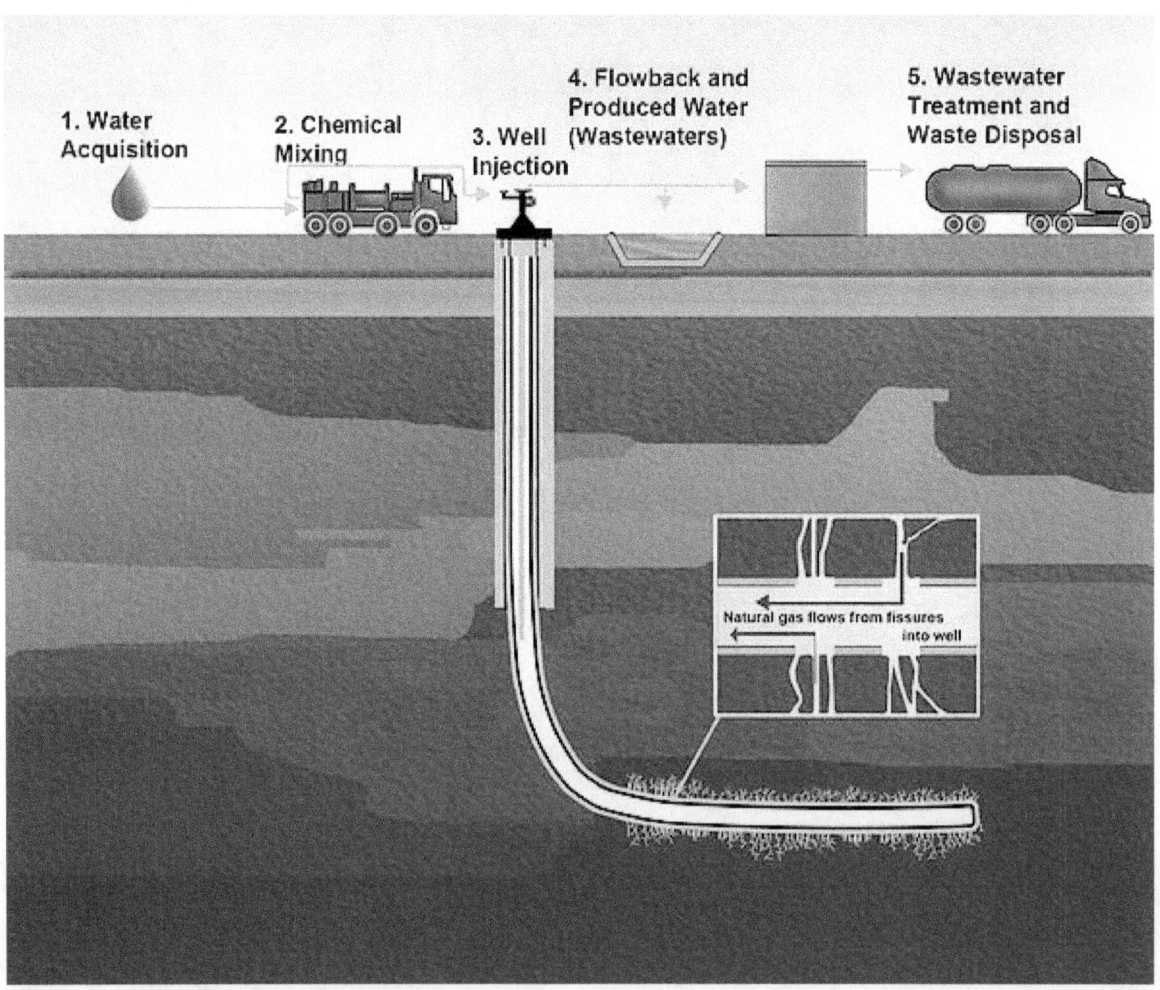

Illustration of hydraulic fracturing and related activities

The **environmental impact of hydraulic fracturing** includes land use and water consumption, methane emissions,[1] air emissions, water contamination, noise pollution, and health effects. Water and air pollution are the biggest risks to human health from hydraulic fracturing. Noise from hydraulic fracturing and associated transport can affect residents and local wildlife; each well pad (in average 10 wells per pad) needs during preparatory and hydraulic fracturing process about 800 to 2,500 days of activity.[2] Research is underway to determine if human health has been affected, and rigorous following of safety procedures and regulation is required to avoid harm and to manage the risk of accidents that could cause harm.[3]

Hydraulic fracturing fluids include proppants and other chemicals. These may include toxic chemicals;[4] In the United States they are allowed to be treated as trade secrets by companies who use them. Lack of knowledge about specific chemicals has complicated efforts to develop risk management policies and to study health effects.[5][6] In other jurisdictions such as the United Kingdom, these chemicals must be made public and are required to be non hazardous in their application.[7]

Water usage by hydraulic fracturing can be a problem in areas that experience water shortage. Surface water may be contaminated through spillage and improperly built and maintained waste pits, in jurisdictions where these are permitted [8] and ground water can be contaminated if the fluid is able to escape. Produced water is managed by underground injection, municipal and commercial wastewater treatment and discharge, self-contained systems at well sites or fields, and recycling to fracture future wells.[9] There is potential for methane to be leaked into the air and ground water. Escape of a methane is a bigger problem in older wells than in ones built under more recent legislation.[2]

Hydraulic fracturing causes induced seismicity called microseismic events or microearthquakes. The magnitude of these events is too small to be detected at the surface, being of magnitude M-3 to M-1 usually. Fluid disposal wells, which are often used in the USA to dispose of polluted waste from several industries, have been responsible for earthquakes up to 5.6M in Oklahoma and other states.[10]

Governments worldwide are developing regulatory frameworks to assess and manage environmental and associated health risks, working under pressure from industry on the one hand, and from anti-fracking groups on the other.[11][12]:3–7 In some countries like France a precautionary approach has been favored and hydraulic fracturing has been banned.[13][14] Some countries such as the United States have adopted the approach of identifying risks before regulating. The United Kingdom's regulatory framework is based on conclusion that the risks associated with hydraulic fracturing are manageable if carried out under effective regulation and if operational best practices are implemented.[11]

9.1   Air emissions

A report for the European Union on the potential risks was produced in 2012. Potential risks are "methane emissions from the wells, diesel fumes and other hazardous pollutants, ozone precursors or odours from hydraulic fracturing equipment, such as compressors, pumps, and valves". Also gases and hydraulic fracturing fluids dissolved in flowback water pose air emissions risks.[2]

"In the UK, all oil and gas operators must minimise the release of gases as a condition of their licence from the Department of Energy and Climate Change (DECC). Natural gas may only be vented for safety reasons." [15]

Also transportation of necessary water volume for hydraulic fracturing, if done by trucks, can cause emissions[16] Using piped water supplies will reduce the number of truck movements necessary.[17]

A report from the Pennsylvania Dept of Environmental Protection indicated that there is little potential for radiation exposure from oil and gas operations.[18]

9.1.1   Climate change

Whether natural gas produced by hydraulic fracturing causes higher well-to-burner emissions than gas produced from conventional wells is a matter of contention. Some studies have found that hydraulic fracturing has higher emissions due to methane released during completing wells as some gas returns to the surface, together with the fracturing fluids. Depending on their treatment, the well-to-burner emissions are 3.5%–12% higher than for conventional gas.[19]

A debate has arisen particularly around a study by professor Robert W. Howarth finding shale gas significantly worse for global warming than oil or coal.[20] Other researchers have criticized Howarth's analysis,[21][22] including Cathles et al., whose estimates were substantially lower."[23] A 2012 industry funded report co-authored by researchers at the United

States Department of Energy's National Renewable Energy Laboratory found emissions from shale gas, when burned for electricity, were "very similar" to those from so-called "conventional well" natural gas, and less than half the emissions of coal.[9]

Several studies which have estimated lifecycle methane leakage from shale gas development and production have found a wide range of leakage rates, from less than 1% of total production to 10%.[24][25][26] According to the Environmental Protection Agency's Greenhouse Gas Inventory a methane leakage rate is about 1.4%.[27] The American Gas Association, an industry trade group, calculated a 1.2% leakage rate.[28] The most comprehensive study of methane leakage from shale gas to date, initiated by the Environmental Defense Fund and released in the Proceedings of the National Academy of Sciences on September 16, 2013, finds that fugitive emissions in key stages of the natural gas production process are significantly lower than estimates in the EPA's national emissions inventory. The study reports direct measurements from 190 onshore natural gas sites, all hydraulically fractured, across the country and estimates a leakage rate of 0.42% for gas production.[24]

9.2   Water consumption

Massive hydraulic fracturing typical of shale wells uses between 1.2 and 3.5 million US gallons (4,500 and 13,200 m^3) of water per well, with large projects using up to 5 million US gallons (19,000 m^3). Additional water is used when wells are refractured.[29][30] An average well requires 3 to 8 million US gallons (11,000 to 30,000 m^3) of water over its lifetime.[30][31][32][33] According to the Oxford Institute for Energy Studies, greater volumes of fracturing fluids are required in Europe, where the shale depths average 1.5 times greater than in the U.S.[34] Whilst the published amounts may seem large, they are small in comparison with the overall water usage in most areas. A study in Texas, which is a water shortage area, indicates "Water use for shale gas is <1% of statewide water withdrawals; however, local impacts vary with water availability and competing demands."[35]

A report by the Royal Society and the Royal Academy of Engineering shows the usage expected for hydraulic fracturing a well is approximately the amount needed to run a 1,000 MW coal-fired power plant for 12 hours.[11] A 2011 report from the Tyndall Centre estimates that to support a 9 billion cubic metres per annum (320×10^9 cu ft/a) gas production industry, between 1.25 to 1.65 million cubic metres (44×10^6 to 58×10^6 cu ft) would be needed annually,[36] which amounts to 0.01% of the total water abstraction nationally.

Concern has been raised over the increasing quantities of water for hydraulic fracturing in areas that experience water stress. Use of water for hydraulic fracturing can divert water from stream flow, water supplies for municipalities and industries such as power generation, as well as recreation and aquatic life.[37] The large volumes of water required for most common hydraulic fracturing methods have raised concerns for arid regions, such as the Karoo in South Africa,[38] and in drought-prone Texas, in North America.[39] It may also require water overland piping from distant sources.[32]

A 2014 life cycle analysis of natural gas electricity by the National Renewable Energy Laboratory concluded that electricity generated by natural gas from massive hydraulically fractured wells consumed between 249 gallons per megawatt-hour (gal/MWhr) (Marcellus trend) and 272 gal/MWhr (Barnett Shale). The water consumption for the gas from massive hydraulic fractured wells was from 52 to 75 gal/MWhr greater (26 percent to 38 percent greater) than the 197 gal/MWhr consumed for electricity from conventional onshore natural gas.[40]

Some producers have developed hydraulic fracturing techniques that could reduce the need for water.[41] Using carbon dioxide, liquid propane or other gases instead of water have been proposed to reduce water consumption.[42] After it is used, the propane returns to its gaseous state and can be collected and reused. In addition to water savings, gas fracturing reportedly produces less damage to rock formations that can impede production.[41] Recycled flowback water can be reused in hydraulic fracturing.[19] It lowers the total amount of water used and reduces the need to dispose of wastewater after use. The technique is relatively expensive, however, since the water must be treated before each reuse and it can shorten the life of some types of equipment.[43]

9.3   Water contamination

9.3.1   Injected fluid

In the United States, hydraulic fracturing fluids include proppants, radionuclide tracers, and other chemicals, many of which are toxic.[4] The type of chemicals used in hydraulic fracturing and their properties vary. While most of them are common and generally harmless, some chemicals are carcinogenic.[4] Out of 2,500 products used as hydraulic fracturing additives in the United States, 652 contained one or more of 29 chemical compounds which are either known or possible human carcinogens, regulated under the Safe Drinking Water Act for their risks to human health, or listed as hazardous air pollutants under the Clean Air Act.[4] Another 2011 study identified 632 chemicals used in United States natural gas operations, of which only 353 are well-described in the scientific literature.[44] The Ground Water Protection Council has launched FracFocus.org, an online voluntary disclosure database for hydraulic fracturing fluids funded by oil and gas trade groups and the Department of Energy.[6][45]

The European Union regulatory regime requires full disclosure of all additives.[5] According to the EU groundwater directive of 2006, "in order to protect the environment as a whole, and human health in particular, detrimental concentrations of harmful pollutants in groundwater must be avoided, prevented or reduced."[46] In the United Kingdom, only chemicals that are "non hazardous in their application" are licensed by the Environment Agency.[7]

Some of the water used in hydraulic fracturing is recovered at the surface as flowback or later production brine. The water left in place is called residual treatment water. According to Engelder and Cathles, this residual treatment water becomes permanently sequestered in the shale and cannot seep into and contaminate ground water.[47]

9.3.2   Flowback

Less than half of injected water is recovered as flowback or later production brine, and in many cases recovery is <30%.[47] As the fracturing fluid flows back through the well, it consists of spent fluids and may contain dissolved constituents such as minerals and brine waters.[48] In some cases, depending on the geology of formation, it may contain uranium, radium, radon and thorium.[49] Estimates of the amount of injected fluid returning to the surface range from 15-20% to 30–70%.[47][48][50]

Approaches to managing these fluids, commonly known as produced water, include underground injection, municipal and commercial wastewater treatment and discharge, self-contained systems at well sites or fields, and recycling to fracture future wells.[9][48][51][52] The vacuum multi-effect membrane distillation system as a more effective treatment system has been proposed for treatment of flowback.[53] However, the quantity of waste water needing treatment and the improper configuration of sewage plants have become an issue in some regions of the United States. Part of the wastewater from hydraulic fracturing operations is processed there by public sewage treatment plants, which are not equipped to remove radioactive material and are not required to test for it.[54][55]

9.3.3   Surface spills

Surface spills related to the hydraulic fracturing occur mainly because of equipment failure or engineering misjudgments.[8]

Volatile chemicals held in waste water evaporation ponds can to evaporate into the atmosphere, or overflow. The runoff can also end up in groundwater systems. Groundwater may become contaminated by trucks carrying hydraulic fracturing chemicals and wastewater if they are involved in accidents on the way to hydraulic fracturing sites or disposal destinations.[56]

In the evolving European Union legislation, it is required that "Member States should ensure that the installation is constructed in a way that prevents possible surface leaks and spills to soil, water or air." [57] Evaporation and open ponds are not permitted. Regulations call for all pollution pathways to be identified and mitigated. The of use chemical proof drilling pads to contain chemical spills is required. In the UK, total gas security is required, and venting of methane is only permitted in an emergency.[58][59][60]

9.3.4 Methane

In September 2014, a study from the US 'Proceedings of the National Academy of Sciences' released a report that indicated that methane contamination can be correlated to distance from a well in wells that were known to leak. This however was not caused by the hydraulic fracturing process, but by poor cementation of casings.[61][62]

Groundwater methane contamination has adverse effect on water quality and in extreme cases may lead to potential explosion.[63] A scientific study conducted by researchers of Duke University found high correlations of gas well drilling activities, including hydraulic fracturing, and methane pollution of the drinking water.[63] According to the 2011 study of the MIT Energy Initiative, "there is evidence of natural gas (methane) migration into freshwater zones in some areas, most likely as a result of substandard well completion practices i.e. poor quality cementing job or bad casing, by a few operators."[64] A 2013 Duke study suggested that either faulty construction (defective cement seals in the upper part of wells, and faulty steel linings within deeper layers) combined with a peculiarity of local geology may be allowing methane to seep into waters; the latter cause may also release injected fluids to the aquifer.[65] Abandoned gas and oil wells also provide conduits to the surface in areas like Pennsylvania, where these are common.[66]

Some drinking water aquifers naturally contain methane, and drawing down the water level in the aquifer may cause an increase of methane in the drinking water, unrelated to oil or gas drilling.[67] Tests can distinguish between the biogenic methane created by bacteria at shallow depths, and the thermogenic methane, which forms under conditions of high pressure and temperature deeper underground. Most oil and gas development produces the deeper-sourced thermogenic methane.[67][68] Although methane that occurs naturally in shallow aquifers is usually biogenic, some drinking-water aquifers contain naturally occurring thermogenic methane,[69] or mixed biogenic-thermogenic methane.[63]

A study by Cabot Oil and Gas examined the Duke study using a larger sample size, found that methane concentrations were related to topography, with the highest readings found in low-lying areas, rather than related to distance from gas production areas. Using a more precise isotopic analysis, they showed that the methane found in the water wells came from both the formations where hydraulic fracturing occurred, and from the shallower formations.[68] The Colorado Oil & Gas Conservation Commission investigates complaints from water well owners, and has found some wells to contain biogenic methane unrelated to oil and gas wells, but others that have thermogenic methane due to oil and gas wells with leaking well casing.[67] A review published in February 2012 found no direct evidence that hydraulic fracturing actual injection phase resulted in contamination of ground water, and suggests that reported problems occur due to leaks in its fluid or waste storage apparatus; the review says that methane in water wells in some areas probably comes from natural resources.[70][71]

Another 2013 review found that hydraulic fracturing technologies are not free from risk of contaminating groundwater, and described the controversy over whether the methane that has been detected in private groundwater wells near hydraulic fracturing sites has been caused by drilling or by natural processes.[72]

9.4 Radionuclides

Main article: Naturally occurring radioactive material

There are naturally occurring radioactive materials (NORM), for example radium, radon,[73] uranium, and thorium,[49][74][75] in shale deposits.[55] Brine co-produced and brought to the surface along with the oil and gas sometimes contains naturally occurring radioactive materials; brine from many shale gas wells, contains these radioactive materials.[55][76][77] When NORM is concentrated or exposed by human activities, such as hydraulic fracturing, EPA classifies it as TENORM (technologically enhanced naturally occurring radioactive material).[78][79]

The U.S. Environmental Protection Agency and regulators in North Dakota consider radioactive material in flowback a potential hazard to workers at hydraulic fracturing drilling and waste disposal sites and those living or working nearby if the correct procedures are not followed.[80][81] A report from the Pennsylvania Department of Environmental Protection indicated that there is little potential for radiation exposure from oil and gas operations.[82]

9.5   Land usage

In the UK, the likely well spacing visualised by the Dec 2013 DECC Strategic Environmental Assessment report indicated that well pad spacings of 5 km were likely in crowded areas, with up to 3 hectares (7.4 acres) per well pad. Each pad could have 24 separate wells. This amounts to 0.16% of land area.[83] A study published in 2015 on the Fayetteville Shale found that a mature gas field impacted about 2% of the land area and substantially increased edge habitat creation. Average land impact per well was 3 hectares (about 7 acres) [84]

9.6   Seismicity

Hydraulic fracturing causes induced seismicity called microseismic events or microearthquakes. These microseismic events are often used to map the horizontal and vertical extent of the fracturing.[85] The magnitude of these events is usually too small to be detected at the surface, although the biggest micro-earthquakes may have the magnitude of about -1.5 (M_w).[86]

9.6.1   Induced seismicity from hydraulic fracturing

As of late 2014, there have been three instances of hydraulic fracturing, through induced seismicity, triggering quakes large enough to be felt by people: one each in the United States, Canada, and England.[87][88] In England, two earthquakes that occurred in April and May 2011 of a magnitude of respectively 1.5 and 2.3 on the Richter scale were felt by local populations. The United Kingdom Department of Energy and Climate Change said the "observed seismicity in April and May 2011 was induced by the hydraulic fracture treatments at Preese Hall", in the North of England.[89][90]

The National Research Council (part of the National Academy of Sciences) has also observed that hydraulic fracturing, when used in shale gas recovery, does not pose a serious risk of causing earthquakes that can be felt.[91]

9.6.2   Induced seismicity from water disposal wells

According to the USGS only a small fraction of roughly 30,000 waste fluid disposal wells for oil and gas operations in the United States have induced earthquakes that are large enough to be of concern to the public.[10] Although the magnitudes of these quakes has been small, the USGS says that there is no guarantee that larger quakes will not occur.[92] In addition, the frequency of the quakes has been increasing. In 2009, there were 50 earthquakes greater than magnitude 3.0 in the area spanning Alabama and Montana, and there were 87 quakes in 2010. In 2011 there were 134 earthquakes in the same area, a sixfold increase over 20th century levels.[93] There are also concerns that quakes may damage underground gas, oil, and water lines and wells that were not designed to withstand earthquakes.[92][94]

Several earthquakes in 2011, including a 4.0 magnitude quake on New Year's Eve that hit Youngstown, Ohio, are likely linked to a disposal of hydraulic fracturing wastewater,[87] according to seismologists at Columbia University.[95] A similar series of small earthquakes occurred in 2012 in Texas. Earthquakes are not common occurrences in either area.[96]

A 2012 US Geological Survey study reported that a "remarkable" increase in the rate of M ≥ 3 earthquakes in the US midcontinent "is currently in progress", having started in 2001 and culminating in a 6-fold increase over 20th century levels in 2011. The overall increase was tied to earthquake increases in a few specific areas: the Raton Basin of southern Colorado (site of coalbed methane activity), and gas-producing areas in central and southern Oklahoma, and central Arkansas.[97] While analysis suggested that the increase is "almost certainly man-made", the USGS noted: "USGS's studies suggest that the actual hydraulic fracturing process is only very rarely the direct cause of felt earthquakes." The increased earthquakes were said to be most likely caused by increased injection of gas-well wastewater into disposal wells.[10] The injection of waste water from oil and gas operations, including from hydraulic fracturing, into saltwater disposal wells may cause bigger low-magnitude tremors, being registered up to 3.3 (M_w).[86]

In 2013, Researchers from Columbia University and the University of Oklahoma demonstrated that in the midwestern United States, some areas with increased human-induced seismicity are susceptible to additional earthquakes triggered by the seismic waves from remote earthquakes. They recommended increased seismic monitoring near fluid injection sites to determine which areas are vulnerable to remote triggering and when injection activity should be ceased.[87][98]

A British Columbia Oil and Gas Commission investigation concluded that a series of 38 earthquakes (magnitudes ranging from 2.2 to 3.8 on the Richter scale) occurring in the Horn River Basin area between 2009 and 2011 were caused by fluid injection during hydraulic fracturing in proximity to pre-existing faults. The tremors were small enough that only one of them was reported felt by people; there were no reports of injury or property damage.[99]

9.7 Noise

Each well pad (in average 10 wells per pad) needs during preparatory and hydraulic fracturing process about 800 to 2,500 days of activity, which may affect residents. In addition, noise is created by transport related to the hydraulic fracturing activities.[2]

The UK Onshore Oil and Gas (UKOOG) is the industry representative body, and it has published a charter that shows how noise concerns will be mitigated, using sound insulation, and heavily silenced rigs where this is needed.[100]

9.8 Safety issues

In July 2013, the United States Federal Railroad Administration listed oil contamination by hydraulic fracturing chemicals as "a possible cause" of corrosion in oil tank cars.[101]

9.9 Health risks

Further information: Environmental impact of hydraulic fracturing in the United States § Health effects

There is worldwide concern over the possible adverse public health implications of hydraulic fracturing activity.[102] A 2013 review on shale gas production in the United States stated, "with increasing numbers of drilling sites, more people are at risk from accidents and exposure to harmful substances used at fractured wells."[103] A 2011 hazard assessment found that most of the chemicals used for hydraulic fracturing and drilling have immediate health effects, and many may have long-term health effects.[104]

In June 2014 Public Health England published a review of the potential public health impacts of exposures to chemical and radioactive pollutants as a result of shale gas extraction in the UK, based on the examination of literature and data from countries where hydraulic fracturing already occurs.[3] The executive summary of the report stated: "An assessment of the currently available evidence indicates that the potential risks to public health from exposure to the emissions associated with shale gas extraction will be low if the operations are properly run and regulated. Most evidence suggests that contamination of groundwater, if it occurs, is most likely to be caused by leakage through the vertical borehole. Contamination of groundwater from the underground hydraulic fracturing process itself (ie the fracturing of the shale) is unlikely. However, surface spills of hydraulic fracturing fluids or wastewater may affect groundwater, and emissions to air also have the potential to impact on health. Where potential risks have been identified in the literature, the reported problems are typically a result of operational failure and a poor regulatory environment."[105]:iii

A 2013 review focusing on Marcellus shale gas hydraulic fracturing and the New York City water supply stated, "Although potential benefits of Marcellus natural gas exploitation are large for transition to a clean energy economy, at present the regulatory framework in New York State is inadequate to prevent potentially irreversible threats to the local environment and New York City water supply. Major investments in state and federal regulatory enforcement will be required to avoid these environmental consequences, and a ban on drilling within the NYC water supply watersheds is appropriate, even if more highly regulated Marcellus gas production is eventually permitted elsewhere in New York State."[106] In 2014, New York State banned hydraulic fracturing entirely, citing health risks.[107]

A 2012 report prepared for the European Union Directorate-General for the Environment identified risks to humans from air pollution and ground water contamination posed by hydraulic fracturing.[2] This led to a series of recommendations in 2014 to mitigate these concerns.[108][109]

A 2012 guidance for pediatric nurses in the US, said that hydraulic fracturing had a potential negative impact on public health, and that pediatric nurses should be prepared to gather information on such topics so as to advocate for improved community health.[110]

9.10   Policy and science

Main article: Regulation of hydraulic fracturing

There are two main approaches to regulation that derive from policy debates about how to manage risk and a corresponding debate about how to assess risk.[12]:3–7

The two main schools of regulation are science-based assessment of risk and the taking of measures to prevent harm from those risks through an approach like hazard analysis, and the precautionary principle, where action is taken before risks are well-identified.[111] The relevance and reliability of risk assessments in communities where hydraulic fracturing occurs has also been debated amongst environmental groups, health scientists, and industry leaders. The risks, to some, are overplayed and the current research is insufficient in showing the link between hydraulic fracturing and adverse health effects, while to others the risks are obvious and risk assessment is underfunded.[112]

Different regulatory approaches have thus emerged. In France and Vermont for instance, a precautionary approach has been favored and hydraulic fracturing has been banned based on two principles: the precautionary principle and the prevention principle.[13][14] Nevertheless, some States such as the U.S. have adopted a risk assessment approach, which had led to many regulatory debates over the issue of hydraulic fracturing and its risks.

In the UK, the regulatory framework is largely being shaped by a report commissioned by the UK Government in 2012, whose purpose was to identify the problems around hydraulic fracturing and to advise the country's regulatory agencies. Jointly published by the Royal Society and the Royal Academy of Engineering, under the chairmanship of Professor Robert Mair, the report features ten recommendations covering issues such as groundwater contamination, well integrity, seismic risk, gas leakages, water management, environmental risks, best practice for risk management, and also includes advice for regulators and research councils.[11][113] The report was notable for stating that the risks associated with hydraulic fracturing are manageable if carried out under effective regulation and if operational best practices are implemented.

A 2013 review concluded that, in the US, confidentiality requirements dictated by legal investigations have impeded peer-reviewed research into environmental impacts.[72]

9.11   See also

- Directional drilling
- Environmental concerns with electricity generation
- Environmental impact of the petroleum industry
- Environmental impact of the oil shale industry

9.12   References

[1] http://www.mintpressnews.com/2500-square-mile-methane-plume-silently-hovering-western-us/200313/

[2] Broomfield 2012

[3] Public Health England. 25 June 2014 PHE-CRCE-009: Review of the potential public health impacts of exposures to chemical and radioactive pollutants as a result of shale gas extraction ISBN 978-0-85951-752-2

[4] Chemicals Used in Hydraulic Fracturing (PDF) (Report). Committee on Energy and Commerce U.S. House of Representatives. April 18, 2011.

[5] Healy 2012

[6] Hass, Benjamin (14 August 2012). "Fracking Hazards Obscured in Failure to Disclose Wells". *Bloomberg News*. Retrieved 27 March 2013.

[7] "Developing Onshore Shale Gas and Oil – Facts about 'Fracking'" (PDF). Department of Energy and Climate Change. Retrieved 14 October 2014.

[8] Walter, Laura (22 May 2013). "AIHce 2013: Investigating Surface Spills in the Fracking Industry". Penton. EHSToday.

[9] Logan, Jeffrey (2012). Natural Gas and the Transformation of the U.S. Energy Sector: Electricity (PDF) (Report). Joint Institute for Strategic Energy Analysis. Retrieved 27 March 2013.

[10] "Man-Made Earthquakes Update". United States Geological Survey. 2014-01-17. Retrieved 2014-03-30.

[11] "Shale gas extraction: Final report". The Royal Society. 29 June 2012. Retrieved 10 October 2014.

[12] Office of Research and Development US Environmental Protection Agency. November 2011 Plan to Study the Potential Impacts of Hydraulic Fracturing on Drinking Water Resources

[13] "LOI n° 2011-835 du 13 juillet 2011 visant à interdire l'exploration et l'exploitation des mines d'hydrocarbures liquides ou gazeux par fracturation hydraulique et à abroger les permis exclusifs de recherches comportant des projets ayant recours à cette technique"

[14] "Vermont Act 152"

[15] UK Department of Energy and Climate Change. February 2014 "Fracking UK shale: local air quality"

[16] Fernandez, John Michael; Gunter, Matthew. "Hydraulic Fracturing: Environmentally Friendly Practices" (PDF). Houston Advanced Research Center. Retrieved 2012-12-29.

[17] "Fracking UK shale: water" (PDF). DECC. Retrieved 13 Nov 2014.

[18] Pennsylvania, Dept of Environmental Protection. "DEP Study Shows There is Little Potential for Radiation Exposure from Oil and Gas Development" (PDF). Pennsylvania DEP. Retrieved Jan 2015.

[19] IEA (2011). *World Energy Outlook 2011*. OECD. pp. 91; 164. ISBN 978 92 64 12413 4.

[20] Howarth, Robert W.; Santoro, Renee; Ingraffea, Anthony (13 March 2011). "Methane and the greenhouse-gas footprint of natural gas from shale formations" (PDF). *Climatic Change* (Springer) **106** (4): 679–690. doi:10.1007/s10584-011-0061-5. Retrieved 2012-05-07.

[21] Cathles, Lawrence M.; Brown, Larry; Taam, Milton; Hunter, Andrew (2011). "A commentary on "The greenhouse-gas footprint of natural gas in shale formations"". *Climatic Change* **113**: 525–535. doi:10.1007/s10584-011-0333-0. Retrieved 7 August 2013.

[22] Stephen Leahy (24 January 2012). "Shale Gas a Bridge to More Global Warming". IPS. Retrieved 4 February 2012.

[23] Howarth, Robert W.; Santoro, Renee; Ingraffea, Anthony (1 February 2012). "Venting and leaking of methane from shale gas development: Response to Cathles et al." (PDF). *Climatic Change* (Springer) **113**: 537–549. doi:10.1007/s10584-012-0401-0. Retrieved 4 February 2012.

[24] Allen, David T.; Torres, Vincent N.; Thomas, James; Sullivan, David W.; Harrison, Matthew; Hendler, Al; Herndon, Scott C.; Kolb, Charles E.; Fraser, Matthew P.; Hill, A. Daniel; Lamb, Brian K.; Miskimins, Jennifer; Sawyer, Robert F.; Seinfeld, John H. (16 September 2013). "Measurements of methane emissions at natural gas production sites in the United States" (PDF). *Proceedings of the National Academy of Sciences* **110**: 17768–17773. doi:10.1073/pnas.1304880110. Retrieved 2013-10-02.

[25] Trembath, Alex; Luke, Max; Shellenberger, Michael; Nordhaus, Ted (June 2013). Coal Killer: How Natural Gas Fuels the Clean Energy Revolution (PDF) (Report). Breakthrough institute. p. 22. Retrieved 2 October 2013.

[26] http://onlinelibrary.wiley.com/doi/10.1002/2014EF000265/full

[27] Bradbury, James; Obeiter, Michael (2013-05-06). "5 Reasons Why It's Still Important To Reduce Fugitive Methane Emissions". World Resources Institute. Retrieved 2013-10-02.

[28] "The Importance of Accurate Data". True Blue Natural Gas. Retrieved 27 March 2013.

[29] Andrews, Anthony; et al. (30 October 2009). Unconventional Gas Shales: Development, Technology, and Policy Issues (PDF) (Report). Congressional Research Service. pp. 7; 23. Retrieved 22 February 2012.

[30] Abdalla, Charles W.; Drohan, Joy R. (2010). Water Withdrawals for Development of Marcellus Shale Gas in Pennsylvania. Introduction to Pennsylvania's Water Resources (PDF) (Report). The Pennsylvania State University. Retrieved 16 September 2012. Hydrofracturing a horizontal Marcellus well may use 4 to 8 million gallons of water, typically within about 1 week. However, based on experiences in other major U.S. shale gas fields, some Marcellus wells may need to be hydrofractured several times over their productive life (typically five to twenty years or more)

[31] GWPC & ALL Consulting 2012

[32] Arthur, J. Daniel; Uretsky, Mike; Wilson, Preston (May 5–6, 2010). *Water Resources and Use for Hydraulic Fracturing in the Marcellus Shale Region* (PDF). Meeting of the American Institute of Professional Geologists. Pittsburgh: ALL Consulting. p. 3. Retrieved 2012-05-09.

[33] Cothren, Jackson. Modeling the Effects of Non-Riparian Surface Water Diversions on Flow Conditions in the Little Red Watershed (PDF) (Report). U. S. Geological Survey, Arkansas Water Science Center Arkansas Water Resources Center, American Water Resources Association, Arkansas State Section Fayetteville Shale Symposium 2012. p. 12. Retrieved 16 September 2012. ...each well requires between 3 and 7 million gallons of water for hydraulic fracturing and the number of wells is expected to grow in the future

[34] Faucon, Benoît (17 September 2012). "Shale-Gas Boom Hits Eastern Europe". WSJ.com. Retrieved 17 September 2012.

[35] Nicot, Jean-Philippe (2 Mar 2012). "Water Use for Shale-Gas Production in Texas, U.S." (PDF). *Environmental Science and Technology*. Retrieved 1 Nov 2014.

[36] Tyndall center report

[37] Upton, John (August 15, 2013). "Fracking company wants to build new pipeline — for water". *Grist*. Retrieved August 16, 2013.

[38] Urbina, Ian (30 December 2011). "Hunt for Gas Hits Fragile Soil, and South Africans Fear Risks". *The New York Times*. Retrieved 23 February 2012. Covering much of the roughly 800 miles between Johannesburg and Cape Town, this arid expanse – its name [Karoo] means "thirsty land" – sees less rain in some parts than the Mojave Desert.

[39] Staff (16 June 2013). "Fracking fuels water battles". *Politico*. Associated Press. Retrieved 26 June 2013.

[40] Life Cycle Analysis of Natural Gas Extraction and Power Generation, NREL, DOE/NETL-2014-1646, 29 May 2014.

[41] "Texas Water Report: Going Deeper for the Solution". Texas Comptroller of Public Accounts. Retrieved 2014-02-11.

[42] Bullis, Kevin (2013-03-22). "Skipping the Water in Fracking". *MIT Technology Review*. Retrieved 2014-03-30.

[43] Sider, Alison; Lefebvre, Ben (20 November 2012). "Drillers Begin Reusing 'Frack Water.' Energy Firms Explore Recycling Options for an Industry That Consumes Water on Pace With Chicago". *The Wall Street Journal*. Retrieved 20 October 2013.

[44] Colborn, Theo; Kwiatkowski, Carol; Schultz, Kim; Bachran, Mary (2011). "Natural Gas Operations from a Public Health Perspective" (PDF). *Human and Ecological Risk Assessment: an International Journal* (Taylor & Francis) **17** (5): 1039–1056. doi:10.1080/10807039.2011.605662.

[45] Soraghan, Mike (13 December 2013). "White House official backs FracFocus as preferred disclosure method". *E&E News*. Retrieved 27 March 2013.

[46] "EU Groundwater directive".

[47] Engelder, Terry; Cathles, Laurence M. (September 2014). "The fate of residual treatment water in gas shale" (PDF). *Journal of Unconventional Oil and Gas Resources* **7**: 33–48. doi:10.1016/j.juogr.2014.03.002.

[48] Arthur, J. Daniel; Langhus, Bruce; Alleman, David (2008). An overview of modern shale gas development in the United States (PDF) (Report). ALL Consulting. p. 21. Retrieved 2012-05-07.

[49] Weinhold, Bob (19 September 2012). "Unknown Quantity: Regulating Radionuclides in Tap Water". *Environmental Health Perspectives*. NIEHS, NIH. Retrieved 11 February 2012. Examples of human activities that may lead to radionuclide exposure include mining, milling, and processing of radioactive substances; wastewater releases from the hydraulic fracturing of oil and natural gas wells... Mining and hydraulic fracturing, or "fracking", can concentrate levels of uranium (as well as radium, radon, and thorium) in wastewater...

[50] Staff. Waste water (Flowback)from Hydraulic Fracturing (PDF) (Report). Ohio Department of Natural Resources. Retrieved 29 June 2013. Most of the water used in fracturing remains thousands of feet underground, however, about 15-20 percent returns to the surface through a steel-cased well bore and is temporarily stored in steel tanks or lined pits. The wastewater which returns to the surface after hydraulic fracturing is called flowback

[51] Hopey, Don (1 March 2011). "Gas drillers recycling more water, using fewer chemicals". *Pittsburgh Post-Gazette*. Retrieved 27 March 2013.

[52] Litvak, Anya (21 August 2012). "Marcellus flowback recycling reaches 90 percent in SWPA.". *Pittsburgh Business Times*. Retrieved 27 March 2013.

[53] "Monitor: Clean that up". *The Economist*. 2013-11-30. Retrieved 2013-12-15.

[54] David Caruso (2011-01-03). "44,000 Barrels of Tainted Water Dumped Into Neshaminy Creek. We're the only state allowing tainted water into our rivers". NBC Philadelphia. Associated Press. Retrieved 2012-04-28.

[55] Urbina, Ian (26 February 2011). "Regulation Lax as Gas Wells' Tainted Water Hits Rivers". *The New York Times*. Retrieved 22 February 2012.

[56] Energy Institute (February 2012). Fact-Based Regulation for Environmental Protection in Shale Gas Development (PDF) (Report). University of Texas at Austin. p. ?. Retrieved 29 February 2012.

[57] "COMMISSION RECOMMENDATION of 22 January 2014 on minimum principles for the exploration and production of hydrocarbons (such as shale gas) using high-volume hydraulic fracturing". EUR-LEX. Retrieved Nov 2014.

[58] European, Commission. "Environmental Aspects on Unconventional Fossil Fuels". Retrieved 27 Oct 2014.

[59] "Fracking UK shale : local air quality" (PDF). DECC. UK Govt. Retrieved 27 Oct 2014.

[60] "Fracking UK shale : water" (PDF). DECC. UK Govt. Retrieved 27 Oct 2014.

[61] abstract

[62] full report

[63] Osborn, Stephen G.; Vengosh, Avner; Warner, Nathaniel R.; Jackson, Robert B. (2011-05-17). "Methane contamination of drinking water accompanying gas-well drilling and hydraulic fracturing" (PDF). *Proceedings of the National Academy of Sciences of the United States of America* **108** (20): 8172–8176. doi:10.1073/pnas.1100682108. Retrieved 2011-10-14.

[64] Moniz, Jacoby & Meggs 2012

[65] Ehrenburg, Rachel (25 June 2013). "News in Brief: High methane in drinking water near fracking sites. Well construction and geology may both play a role". Science News. Retrieved 26 June 2013.

[66] Detrow, Scott (9 October 2012). "Perilous Pathways: How Drilling Near An Abandoned Well Produced a Methane Geyser". *StateImpact Pennsylvania*. NPR. Retrieved 29 June 2013.

[67] "Gasland Correction Document" (PDF). Colorado Oil & Gas Conservation Commission. Retrieved 7 August 2013.

[68] Molofsky, L. J.; Connor, J. A.; Shahla, K. F.; Wylie, A. S.; Wagner, T. (December 5, 2011). "Methane in Pennsylvania Water Wells Unrelated to Marcellus Shale Fracturing". *Oil and Gas Journal* (Pennwell Corporation) **109** (49): 54–67. (subscription required).

[69] Texas Railroad Commission, Press release: staff report on Parker, Texas, 22 Mar. 2011.

[70] "Fracking Acquitted of Contaminating Groundwater". *Science* **335**: 898. 24 February 2012. doi:10.1126/science.335.6071.898.

[71] Erik Stokstad (16 February 2012). "Mixed Verdict on Fracking". *Science Now*.

[72] Vidic, R.D.; et al. (May 17, 2013). "Impact of Shale Gas Development on Regional Water Quality" (PDF). *Science* **340** (1235009): 826. doi:10.1126/science.1235009. PMID 23687049. Retrieved 29 September 2014.

[73] Staff. "Radon in Drinking Water: Questions and Answers" (PDF). US Environmental Protection Agency. Retrieved 7 August 2012.

[74] Heather Smith (7 March 2013). "County's potential for fracking is undetermined". *Environment / Pollution*. Discover Magazine. Retrieved 11 August 2013.

[75] Lubber, Mindy (28 May 2013). "Escalating Water Strains In Fracking Regions". Forbes. Retrieved 20 October 2013.

[76] Linda Marsa (1 August 2011). "Fracking Nation. Environmental concerns over a controversial mining method could put America's largest reservoirs of clean-burning natural gas beyond reach. Is there a better way to drill?". *Environment / Pollution*. Discover Magazine. Retrieved 5 August 2011.

[77] White, Jeremy; Park, Haeyoun; Urbina, Ian; Palmer, Griff (26 February 2011). "Toxic Contamination From Natural Gas Wells". *The New York Times*.

[78] "TENORM Sources". United States Environmental Protection Agency. Retrieved 2012-09-12.

[79] "Oil and Gas Production Wastes". United States Environmental Protection Agency. Retrieved 2012-09-12.

[80] "Radioactive Waste from Oil and Gas Drilling" (PDF). United States Environmental Protection Agency. April 2006. Retrieved 2013-08-11.

[81] McMahon, Jeff (24 July 2013). "Strange Byproduct Of Fracking Boom: Radioactive Socks". Forbes. Retrieved 28 July 2013.

[82] Pennsylvania, Dept of Environmental Protection. "DEP Study Shows There is Little Potential for Radiation Exposure from Oil and Gas Development" (PDF). Pensylvania DEP. Retrieved Jan 2015.

[83] "Strategic Environmental Assessment for Further Onshore Oil and Gas Licensing" (PDF). Department of Energy and Climate Change. June 2014. p. ?. Retrieved 11 November 2014.

[84] http://link.springer.com/article/10.1007/s00267-014-0440-6

[85] Bennet, Les; et al. "The Source for Hydraulic Fracture Characterization" (PDF). *Oilfield Review* (Schlumberger) (Winter 2005/2006): 42–57. Retrieved 2012-09-30.

[86] Zoback, Kitasei & Copithorne 2010

[87] Kim, Won-Young 'Induced seismicity associated with fluid injection into a deep well in Youngstown, Ohio', Journal of Geophysical Research-Solid Earth

[88] Begley, Sharon; McAllister, Edward (12 July 2013). "News in Science: Earthquakes may trigger fracking tremors". *ABC Science* (Reuters). Retrieved 17 December 2013.

[89] Dr Christopher A. GREEN, Professor Peter STYLES. "PREESE HALL SHALE GAS FRACTURING REVIEW & RECOMMENDAT IONS FOR INDUCED SEISMIC MITI GATION" (PDF). DECC. Retrieved Nov 2014.

[90] de Pater, C.J.; Baisch, S. (2 November 2011). Geomechanical Study of Bowland Shale Seismicity (PDF) (Report). Cuadrilla Resources. Retrieved 22 February 2012.

[91] "Induced Seismicity Potential in Energy Technologies". National Academies Press. Retrieved 27 March 2013. The process of hydraulic fracturing a well as presently implemented for shale gas recovery does not pose a high risk for inducing felt seismic events.

[92] Rachel Maddow, Terrence Henry (7 August 2012). *Rachel Maddow Show: Fracking waste messes with Texas* (video). MSNBC. Event occurs at 9:24 - 10:35.

[93] Soraghan, Mike (29 March 2012). "'Remarkable' spate of man-made quakes linked to drilling, USGS team says". *EnergyWire* (E&E). Retrieved 2012-11-09.

[94] Henry, Terrence (6 August 2012). "How Fracking Disposal Wells Are Causing Earthquakes in Dallas-Fort Worth". *State Impact Texas*. NPR. Retrieved 9 November 2012.

[95] "Ohio Quakes Probably Triggered by Waste Disposal Well, Say Seismologists" (Press release). Lamont–Doherty Earth Observatory. 6 January 2012. Retrieved 22 February 2012.

[96] "EPA Underground Injection Control Program". Retrieved 2012-04-13.

[97] Ellsworth, W. L.; Hickman, S.H.; McGarr, A.; Michael, A. J.; Rubinstein, J. L. (18 April 2012). *Are seismicity rate changes in the midcontinent natural or manmade?*. Seismological Society of America 2012 meeting. San Diego, California: Seismological Society of America. Retrieved 2014-02-23.

[98] van der Elst1, Nicholas J.; Savage, Heather M.; Keranen, Katie M; Abers, Geoffrey A. (12 July 2013). "Enhanced Remote Earthquake Triggering at Fluid-Injection Sites in the Midwestern United States". *Science* (ACS Publications). 341 (6142): 164–167. doi:10.1126/science.1238948. PMID 23846900.

[99] "Fracking causes minor earthquakes, B.C. regulator says". *The Canadian Press* (Canadian Broadcast Company — British Columbia). 6 September 2012. Retrieved 2012-10-28.

[100] "What it looks like Noise chapter". UKOOG. Retrieved 11 Nov 2014.

[101] Frederick J. Herrmann, Federal Railroad Administration, letter to American Petroleum Institute, 17 July 2013, p.4.

[102] Finkel ML, Hays J (October 2013). "The implications of unconventional drilling for natural gas: a global public health concern". *Public Health* (Review) **127** (10): 889–93. doi:10.1016/j.puhe.2013.07.005. PMID 24119661. This *in silico* epidemiologic study will analyse at 2.6 million electronic health records of patients in 31 Pennsylvania counties for respiratory, cardiovascular, cerebrovascular, and pregnancy outcomes.

[103] Centner, Terence J. (September 2013). "Oversight of shale gas production in the United States and the disclosure of toxic substances". *Resources Policy* **38** (3): 233–240. doi:10.1016/j.resourpol.2013.03.001. Retrieved 29 September 2014.

[104] Colborn, Theo; et al. (September 20, 2011). "Natural Gas Operations from a Public Health Perspective" (PDF). *Human and Ecological Risk Assessment* **17** (5): 1039–1056. doi:10.1080/10807039.2011.605662.

[105] A. Kibble, T. Cabianca, Z. Daraktchieva, T. Gooding, J. Smithard, G. Kowalczyk, N. P. McColl, M. Singh, L. Mitchem, P. Lamb, S. Vardoulakis and R. Kamanyire (January 2014). Review of the Potential Public Health Impacts of Exposures to Chemical and Radioactive Pollutants as a Result of the Shale Gas Extraction Process (PDF) (Report). Public Health England. PHE-CRCE-009.

[106] Eaton TT. Science-based decision-making on complex issues: Marcellus shale gas hydrofracking and New York City water supply. Sci Total Environ. 2013 Sep 1;461-462:158-69. doi: 10.1016/j.scitotenv.2013.04.093. Epub 2013 May 28. PMID 23722091

[107] Brady, Jeff (December 18, 2014). "Citing Health, Environment Concerns, New York Moves To Ban Fracking". *National Public Radio*. Retrieved 6 January 2015.

[108] "EU Commission minimum principles for the exploration and production of hydrocarbons (such as shale gas) using high-volume hydraulic fracturing". EUR LEX. Retrieved Nov 2014.

[109] "Energy and environment". EUR LEX.

[110] Lauver LS (August 2012). "Environmental health advocacy: an overview of natural gas drilling in northeast Pennsylvania and implications for pediatric nursing". *J Pediatr Nurs* **27** (4): 383–9. doi:10.1016/j.pedn.2011.07.012. PMID 22703686.

[111] Editors, ParisTech Review March 28th, 2014 Is it really possible to enforce the precautionary principle?

[112] Williams, Laurence, John "Framing fracking: public responses to potential unconventional fossil fuel exploitation in the North of England", Durham thesis, Durham University, 2014

[113] Royal Society 2012

9.13 Bibliography

- Broomfield, Mark (2012-08-10). Support to the identification of potential risks for the environment and human health arising from hydrocarbons operations involving hydraulic fracturing in Europe (PDF) (Report) (17c). European Commission. pp. vi–xvi. ED57281. Retrieved 2014-09-29.

- Brown, Valerie J. (February 2007). "Industry Issues: Putting the Heat on Gas". *Environmental Health Perspectives* (US National Institute of Environmental Health Sciences) **115** (2): A76. doi:10.1289/ehp.115-a76. PMC 1817691. PMID 17384744. Retrieved 2012-05-01.

- Ground Water Protection Council; ALL Consulting (April 2009). Modern Shale Gas Development in the United States: A Primer (PDF) (Report). DOE Office of Fossil Energy and National Energy Technology Laboratory. pp. 56–66. DE-FG26-04NT15455. Retrieved 24 February 2012.

- Healy, Dave (July 2012). Hydraulic Fracturing or 'Fracking': A Short Summary of Current Knowledge and Potential Environmental Impacts (PDF) (Report). Environmental Protection Agency. Retrieved 28 July 2013.

- Jenner, Steffen; Lamadrid, Alberto J. (2013). "Shale gas vs. coal: Policy implications from environmental impact comparisons of shale gas, conventional gas, and coal on air, water, and land in the United States" (PDF). *Energy Policy* (Elsevier) **53** (53): 442–453. doi:10.1016/j.enpol.2012.11.010. Retrieved 2014-09-28.

- Mair (Chair), Robert (June 2012). Shale gas extraction in the UK: A review of hydraulic fracturing (PDF) (Report). The Royal Society and the Royal Academy of Engineering. Retrieved 10 October 2014.

- Moniz (chair), Ernest J.; Jacoby (Co-Chair), Henry D.; Meggs (Co-Chair), Anthony J. M. (June 2011). *The future of natural gas: An interdisciplinary MIT study* (PDF). Massachusetts Institute of Technology. Retrieved 8 October 2014.

- Zoback, Mark; Kitasei, Saya; Copithorne, Brad (July 2010). Addressing the Environmental Risks from Shale Gas Development (PDF) (Report). Worldwatch Institute. p. 9. Retrieved 2012-05-24.

Chapter 10

Exemptions for hydraulic fracturing under United States federal law

There are many **exemptions for hydraulic fracturing under United States federal law**: the oil and gas industries are exempt or excluded from certain sections of a number of the major federal environmental laws. These laws range from protecting clean water and air, to preventing the release of toxic substances and chemicals into the environment: the Clean Air Act, Clean Water Act, Safe Drinking Water Act, National Environmental Policy Act, Resource Conservation and Recovery Act, Emergency Planning and Community Right-to-Know Act, and the Comprehensive Environmental Response, Compensation, and Liability Act, commonly known as Superfund.

10.1 Hydraulic fracturing: background

Main article: Hydraulic fracturing

Hydraulic fracturing, also known as fracking, is a process used to extract oil and natural gas. The process to extract oil and natural gas begins with thousands of gallons of water, mixed with a slurry of chemicals, some of which are undisclosed. This liquid mixture is then forced into well casings under high pressure, and then is horizontally injected into bedrock to create cracks or fissures. The forced change in geologic structure allows gas molecules to escape, therefore allowing the natural gas to be harvested.

Hydraulic fracturing has changed the energy scene as a result of many technological advances. Fracking uses both historically-known vertical and horizontal drilling techniques which are used in tandem to extract oil and gas. This process can occur at depths over 10,000 feet deep.

The primary product of hydraulic fracturing is natural gas which consists mostly of methane.[1]

10.2 Clean Water Act

The Clean Water Act is a result of the 1972 amendments to the Federal Water Pollution Control Act, which was passed to ultimately eliminate pollution discharge into any body of water in the United States.[2] One of the major mechanisms for implementing this statute was to create a permitting process for all discharging methods that involved dumping pollutants into streams, lakes, rivers, wetlands, or creeks. The National Pollution Discharge Elimination System (NPDES) permitting requirements apply to all phases of the petroleum industry. Petroleum industry waste, including frac flowback and produced water, cannot be discharged to the waters of the United States, except under an NPDES or equivalent state permit.[3]

In 1987, congress amended the act, requiring the EPA to develop a permitting program for storm water runoff, but the exploration, production, and processing of oil and gas was exempt. The Energy Policy Act of 2005 expanded the

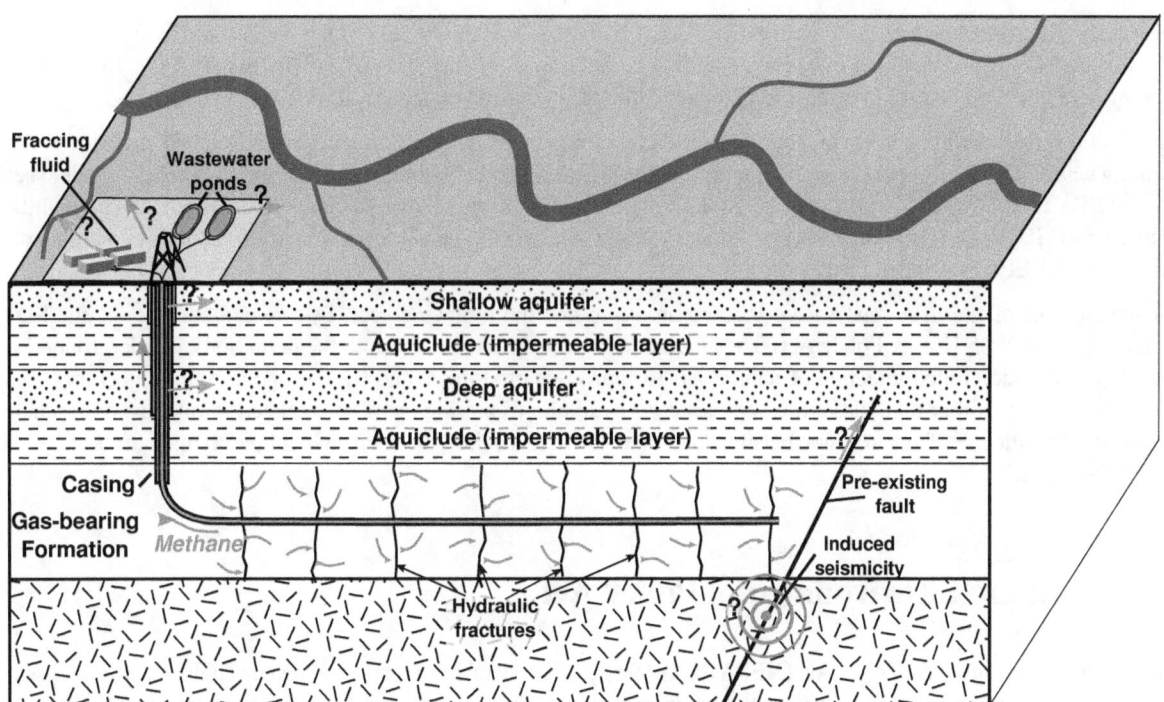

The process of hydraulic fracturing

exemption to include exemptions for runoff from gas and oil construction activities which include "oil and gas exploration, production, process, or treatment operations and transmission facilities."[4]

In 2006 EPA promulgated regulations that would not require oil and gas facilities to obtain storm water runoff permits, if the runoff is "composed entirely of storm water" (40 CFR § 122.26(c)(1)(iii)), which is defined as composed of "precipitation runoff" and "not contaminated by contact with or that has not come into contact with, any overburden, raw material, intermediate products, finished product, byproduct or waste products located on the site of such operations." (40 CFR § 122.26(e)(8)).[5] Any discharges containing other than precipitation runoff, such as petroleum or produced wastewater are still subject to criminal prosecution under the Clean Water Act.[6] EPA's 2006 rule was vacated by the United States Court of Appeals for the 9th Circuit. *NRDC v. EPA*, 526 F.3d 591, 596, 607-08 (2008).

10.3   Safe Drinking Water Act

In 1974, the Safe Drinking Water Act (SDWA) was passed to protect the quality of U.S. public drinking water and aims to protect above and below ground water sources that are or could potentially be used for human consumption.[7] Section C of the SDWA requires the EPA to establish minimum regulations for State Underground Injection Control Programs. Under part C, Section 1421 of the SDWA, underground injection is "the subsurface emplacement of fluids by well injection." The oil and gas industry makes extensive use of Class II injection wells, which are regulated under the SDWA. There are currently about 144,000 such wells with permits issued to SDWA standards. Most Class II injector wells are for enhanced oil recovery, such as waterfloods. About 20 percent of Class II wells are used in waste disposal, to dispose of produced water, usually brine, into deep formations below the base of fresh water.[8]

From the time of the passage of the Safe Drinking Water Act in 1974, the EPA declined to require Class II underground injection permits for hydraulic fracturing. The agency maintained that it was not required to do so, because underground injection was not the "principal function" of the wells. The EPA also cited the "endangerment clause" in SWDA section 1421(b)(2), which directs the EPA to establish regulations which: "... are essential to assure that underground sources of drinking water will not be endangered by such injection." The agency stated that it did not consider hydraulic fracturing to be an endangerment to underground drinking water sources. [9]

The policy was overturned in 1997 by the U.S. Court of Appeals 11th Circuit, which ruled that "hydraulic fracturing activities constitute underground injection according to Section C of the SDWA[10] This required the EPA and state underground injection control programs to regulate hydraulic fracturing under the SDWA.

The EPA responded with a study of potential and actual impacts of hydraulic fracturing of coalbed methane wells on drinking water. The draft report came out for comments in 2002, and the final report came out in 2004.[11] Under section 7.4, the EPA report "concluded that the injection of hydraulic fracturing fluids into coalbed methane wells poses little or no threat to USDWs and does not justify additional study at this time." The exception was for frac fluids containing diesel fuel, which the EPA concluded could pose a threat.[12][13][14]

The conclusions of the EPA report were incorporated into law the following year, by two amendments of the SDWA contained in the 2005 Energy Policy Act. The amendments added two exclusions to the definition of underground injection: ""(i) the underground injection of natural gas for purposes of storage; and (ii) the underground injection of fluids or propping agents (other than diesel fuels) pursuant to hydraulic fracturing operations related to oil, gas, or geothermal production activities.[15] This provision became known to its critics as the "Halliburton loophole" named after oil-services firm Halliburton.

10.4 National Environmental Policy Act

The National Environmental Policy Act (NEPA) of 1969 requires federal agencies to conduct an environmental assessment for all major actions potentially affecting the environment. If the assessment determines that the federal action may significantly alter the environment, then an environmental impact statement is required.[16]

The Energy Policy Act of 2005 created a rebuttable presumption that certain oil and gas related activities authorized by the U.S. Department of the Interior in managing public lands, and the U.S. Department of Agriculture in managing National Forest System Lands are subject to a "categorical exclusion" under NEPA, and do not require an EIS, unless it can be demonstrated that they pose a risk to the environment.[12][17] Congress specified five circumstances for which there would be such a rebuttable presumption that an additional EIS is not required:

"1. Individual surface disturbances of less than five (5) acres so long as the total surface disturbance on the lease is not greater than 150 acres and site-specific analysis in a document prepared pursuant to NEPA has been previously completed.

"2. Drilling an oil and gas well at a location or well pad site at which drilling has occurred previously within five (5) years prior to the date of spudding the well.

"3. Drilling an oil or gas well within a developed field for which an approved land use plan or any environmental document prepared pursuant to NEPA analyzed such drilling as a reasonably foreseeable activity, so long as such plan or document was approved within five (5) years prior to the date of spudding the well

"4. Placement of a pipeline in an approved right-of-way corridor, so long as the corridor was approved within five (5) years prior to the date of placement of the pipeline.

"5. Maintenance of a minor activity, other than any construction or major renovation o(f) a building or facility."[18]

Other than for the exclusions listed above, federal agencies are required by NEPA to do Environmental Impact Statements to evaluate any oil and gas activities which have the potential to seriously affect the environment. Such EIS's are routinely done for specific areas by the Forest Service,[19] the Bureau of Land Management,[20] and the Bureau of Ocean Energy Management.[21]

10.5   Resource Conservation and Recovery Act

The Resource Conservation and Recovery Act (RCRA) of 1976 was passed "to protect human health and the environment from the potential hazards of waste disposal, to conserve energy and natural resources, to reduce the amount of waste generated, and to ensure that wastes are managed in an environmentally sound manner."[22] Subtitle C of RCRA gives the EPA the authority to regulate the generation, transport, treatment, storage and disposal of all deemed hazardous waste.

In December 1978, the EPA issued its proposed RCRA regulations. For RCRA Subtitle C (hazardous waste management), the EPA defined six categories of "special wastes," which were generated in high volumes and were believed to be less hazardous than the other wastes for which RCRA Subtitle C was designed. Among the special wastes were included cement kiln dust, fly ash, mining wastes, and wastes from oil, gas, and geothermal exploration and production. The oil, gas, and geothermal wastes included drilling fluids, produced waters, and other wastes associated with oil and natural gas exploration, development, or production.[23] The EPA proposed that regulation of special wastes under Subtitle C, be deferred until further study.[24]

Prior to the completion of the EPA's regulatory determination, Congress enacted the Solid Waste Disposal Act in 1980 which exempted oil field wastes under section C of RCRA unless the EPA determined that the waste was hazardous.[25]

Each of the six special waste categories was the subject of separate EPA study. In July 1988, the EPA finished its study of oil, gas, and geothermal production wastes, in which it concluded that they did not warrant regulation under RCRA Subtitle C, but noted that they would continue to be regulated under Subtitle D (solid waste disposal). The EPA's decision was based on its determinations that oil, gas, and geothermal production was already regulated by the states, that Subtitle C did not have the regulatory flexibility to deal effectively with the wastes, and that the permitting requirements of Subtitle C would impose unreasonable delays on oil, gas, and geothermal extraction. However, the EPA report identified regulatory gaps for oil and gas wastes, for which it recommended additional rules under existing EPA regulatory authority, under RCRA Subtitle D, the Clean Water Act, and the Safe Water Drinking Act.[26]

Federal regulation of the storage of petroleum was established by the Oil Pollution Act of 1990.[27]

10.6   Emergency Planning and Community Right-to-Know Act

The Emergency Planning and Community Right-to-Know Act, or EPCRA was passed by congress in 1986 and it was created to help communities plan for emergencies that involve hazardous substance spills or releases. The Act requires federal, state, local governments and Indian tribes to inform the public of hazardous and toxic chemicals being used or stored at facilities, their use, and any release into the environment.[28] The provisions of the EPCRA include emergency planning (Sections 301-303),[29] and emergency release notification (Section 304).[30]

The Toxics Release Inventory Reporting (Section 313) of "EPCRA requires the EPA and States to collect data on releases and transfers of listed toxic chemicals." The facilities required to report releases and transfers under section 313 are those in certain industries on a Standard Industrial Classification list determined by the EPA. The EPA has steadily expanded the law's coverage by adding new industrial classifications to the list. As of 2014, the oil and gas industry has not been added to the list, and is therefore exempt from the EPCRA Section 313.[12]

10.7   Comprehensive Environmental Response, Compensation, and Liability Act (Superfund)

The Comprehensive Environmental Response, Compensation, and Liability Act, also known as Superfund was enacted in 1980 to hold all polluters and potentially responsible parties liable for toxic or hazardous substances dumped into the environment. This federal law can be retroactively implemented, and clean-up methods are required for hazardous waste sites under Superfund hold all polluting industries responsible for the costs. As of March 26, 2015, there have been a total of 1,709 Superfund sites, of which 386 (23%) have been remediated.[31]

Under Section 9601(14) of CERCLA, hazardous waste definitions exclude crude petroleum, including crude oil, natural gas liquids, and any of their component fractions. Included in the exemption are refined petroleum products, such as gasoline and diesel fuel, insofar as their content of naturally occurring petroleum compounds. If any spills that would be otherwise classified under the Superfund contain only petroleum compounds, they are exempt from the cleanup process associated with CERCLA.[32] The petroleum exemption does not extend to hazardous contaminants such as PCBs or pesticides, which are sometimes mixed with petroleum product. "Moreover, if the petroleum product and an added hazardous substance are so commingled that, as a practical matter, they cannot be separated, then the entire oil spill is subject to CERCLA response authority." As of 1987, there were at least 153 CERCLA Superfund sites that included waste oil.[33]

Despite the petroleum exemption, the EPA has exercised its power under CERCLA to intervene where it considers oil and gas operations to pose "imminent and substantial danger to the public health or welfare." Citing its CERCLA authority, the EPA has investigated instances of groundwater pollution it believed were related to oil and gas wells, including those at Pavillion, Wyoming, Dimock, Pennsylvania, and a Marcellus shale gas well in Bradford County, Pennsylvania.[34]

Congress addressed petroleum contamination in the 1986 Superfund Amendments and Reauthorization Act, which authorized the EPA to enforce the environmental cleanup of petroleum hydrocarbons released from underground storage tanks. The act also established the Leaking Underground Storage Tank Trust Fund, to fund cleanup of petroleum hydrocarbon released from underground storage tanks at places such as gasoline stations.[35] Sites contaminated by petroleum from leaking underground storage tanks are much more widespread and numerous than CERCLA Superfund sites. The EPA notes that nearly every community has petroleum contamination beneath present or former gasoline stations.[36] As of September 2014, the federally financed but mostly state-run leaking underground storage tank program has found 521,271 petroleum releases from underground storage tanks at 205,000 facilities, 86% of which have been remediated. In fiscal year 2014, 6,847 new leaking tanks were discovered.[37] The program is financed by a federal 0.1-cent tax on petroleum products.[38]

10.8 Debates Surrounding Regulatory Exemptions

There have many debates surrounding the regulatory exemptions for hydraulic fracturing. It has been noted that if not for the exemption for hydraulic fracturing in the Energy Policy Act of 2005 or the RCRA exemption that exempts oil and gas waste from being designated as a hazardous waste, underground injection would have included fracking operations, and the EPA would have had the power to further regulate it as well as enforcing disclosure requirements.[39]

On the other side of this, the oil and gas industry, Congress, and some environmental groups support the idea that states, with greater knowledge about the local economic and ecological landscape, should control the regulatory specificities of fracking.[40] Some contend that these exemptions are carefully analyzed, for example, an EPA study revealed that fracking injection "posed little or no threat to drinking water,"[41] and some still contend that there is a lack of scientific evidence to prove otherwise. Many industry leaders contend that the regulations currently in place are sufficient, and as demonstrated by the current regulatory framework, the majority of members in Congress believe that these regulations, and their exemptions, are sufficient to ensure the safety and protect the health of the public and the environment.

10.9 References

[1] Muller, Richard (2012). *Energy For Future Presidents*. 500 Fifth Avenue, New York, NY 10110: WW. Norton & Company. pp. 87–91. ISBN 978-0-393-08161-9.

[2] "History - Clean Water Act". EPA. Retrieved 22 April 2013.

[3] US EPA, Unconventional extraction in the oil and gas industry, 14 Nov. 2014.

[4] "Environmental Defense Center: Fracking". Retrieved 22 April 2013.

[5] US EPA, Regulation of Oil and Gas Construction Activities, 6 Dec. 2012.

[6] US Attorney's Office, Southern District of Ohio, Monroe County company owner pleads guilty to discharging wastewater into a tributary of the Little Muskingum River, 12 Feb. 2013.

[7] "Safe Drinking Water Act (SDWA)". *Regulatory information*. EPA. Retrieved 20 April 2013.

[8]   Underground Injection Control, accessed 4 February 2015.

[9]   Adam Vann and others, Hydraulic Fracturing: Selected Legal Issues, US Library of Congress, Congressional Research Service, 26 Sept. 2014., p.1-2.

[10] "Bill Text 112th Congress (2011-2012) H.R.1084.IH". Retrieved 23 April 2013.

[11] US EPA, Evaluation of impacts to underground sources of drinking water by hydraulic fracturing of coalbed methane reservoirs, June 2004.

[12] Brady, William J. "Hydraulic Fracturing Regulation in the United States: The Laissez-Faire Approach of the Federal Government and Varying State Regulations" (PDF). Retrieved 20 April 2013.

[13] "Class II Wells - Oil and Gas Related Injection Wells (Class II)". EPA. Retrieved 20 April 2013.

[14] "Evaluation of Impacts to Underground Sources of Drinking Water by Hydraulic Fracturing of Coalbed Methane Reservoirs; National Study Final Report" (PDF). EPA. Retrieved 23 April 2013.

[15] "ENERGY POLICY ACT OF 2005" (PDF). *Authentic Government Information GPO*. Retrieved 23 April 2013.

[16] "National Environmental Policy Act". EPA. Retrieved 23 April 2013.

[17] "42 USC § 15942 - NEPA review". Cornell University. Retrieved 23 April 2013.

[18] US Forest Service, Energy Policy Act of 2005, Use of Section 390 Categorical Exclusions for Oil and Gas Activities, 9 June 2010.

[19] US Forest Service, EIS, Pike and San Isabel national forests, accessed 8 December 2014.

[20] US Bureau of Land Management, Oil and gas strategy, accessed 8 Dec. 2014

[21] US BOEM, Arctic Ocean EIS, accessed 8 Dec. 2014.

[22] "RCRA Corrective Action". EPA. Retrieved 23 April 2013.

[23] "Hazardous Waste; RCRA Subtitle C". EPA. Retrieved 23 April 2013.

[24] US EPA, Special wastes, 15 Nov. 2012.

[25] Earthworks. "The Oil and Gas Industry's Exclusions and Exemptions to Major Statutes" (PDF). www.earthworksaction.org. Retrieved 20 April 2013.

[26] EPA, Regulatory Determination for Oil and Gas and Geothermal Exploration, Development and Production Wastes, 6 July 1988.

[27] US EPA, Summary of the Oil pollution Act, accessed 8 Dec. 2014.

[28] "Emergency Planning and Community Right-to-Know Act (EPCRA) Requirements". US EPA. Retrieved 2 May 2013.

[29] "Emergency Planning and Community Right-to-Know Act (EPCRA) Local Emergency Planning Requirements". US EPA. Retrieved 2 May 2013.

[30] "Emergency Planning and Community Right-to-Know Act (EPCRA) Emergency Release Notification Requirements". US EPA. Retrieved 2 May 2013.

[31] US EPA, [massachusetts institute of technology National Priorities List], accessed 5 Apr. 2015.

[32] US EPA, Specific Substances Excluded Under the CERCLA Petroleum Exemption, 4 Nov. 2014.

[33] US EPA, Scope of the CERCLA Petroleum Exclusion Under Sections lOl(14) and 104(a)(2). 31 July 1987, p.3, 11.

[34] Government Accountability Office, Unconventional Oil and Gas Development, Sept. 2012, p.176-178.

[35] US EPA, What is the history of the underground storage tank program?, accessed 21 Jan. 2015.

[36] US EPAS, UST Program 30th Anniversary, 6 Nov. 2014.

[37] US EPA, Semiannual Report of UST Performance Measures, Fiscal Year 2014, Nov. 2014.

[38] US EPA, Leaking Underground Storage Tank Trust Fund, 6 Mar. 2015.

[39] Kiparsky, Michael; et al. "Regulation of Hydraulic Fracturing in California: A Wastewater and Water Quality Perspective" (PDF). Berkeley Law, University of California Center for Law, Energy, and the Environment. April 2013.

[40] Negro, Sorrell E. "Fracking Wars: Federal, State, and Local Conflicts over the Regulation of Natural Gas Activities." (PDF). Zoning and Planning Law Report. Vol. 35. No. 2. February 2012.

[41] Burger, Michael. "Response, Fracking and Federalism Choice" (PDF). 161 U. PA. L. REV. ONLINE 150 (2013).

Chapter 11

ExxonMobil Electrofrac

ExxonMobil Electrofrac is an *in situ* shale oil extraction technology proposed by ExxonMobil for converting kerogen in oil shale to shale oil.

11.1 Technology

ExxonMobil Electrofrac uses a series of hydraulic fractures created in the oil shale formation. Preferably these fractures should be longitudinal vertical fractures created from horizontal wells and conducting electricity from the heel to the toe of each heating well. For conductivity, an electrically-conductive material such as calcined petroleum coke is injected into the wells in fractures, forming a heating element.[1][2][3] Heating wells are placed in a parallel row with a second horizontal well intersecting them at their toe. This allows opposing electrical charges to be applied at either end. Laboratory experiments have demonstrated that electrical continuity is unaffected by kerogen conversion and that hydrocarbons are expelled from heated oil shale even under *in situ* stress.[3] Planar heaters should be used because they require fewer wells than wellbore heaters and offer a reduced surface footprint. The shale oil is extracted by separate dedicated production wells.[1][2]

11.2 See also

- Shell in situ conversion process
- Chevron CRUSH

11.3 References

[1] *Secure Fuels from Domestic Resources: The Continuing Evolution of America's Oil Shale and Tar Sands Industries* (PDF) (4th ed.). United States Department of Energy. 2010. pp. 38–39. Retrieved 2011-05-06.

[2] Plunkett, Jack W. (2008). *Plunkett's Energy Industry Almanac 2009: The Only Comprehensive Guide to the Energy & Utilities Industry*. Plunkett Research, Ltd. p. 186. ISBN 978-1-59392-128-6. Retrieved 2009-03-14.

[3] Symington, William A.; Olgaard, David L.; Otten, Glenn A.; Phillips, Tom C.; Thomas,Michele M.; Yeakel, Jesse D. (2008-04-20). *ExxonMobil's Electrofrac Process for In Situ Oil Shale Conversion* (PDF). AAAPG Annual Convention. San Antonio: American Association of Petroleum Geologists. Retrieved 2009-04-12.

Chapter 12

Fracking hose

Fracking hoses are used in hydraulic fracturing (fracking). Hydraulic fracturing uses between 1.2 and 3.5 million US gallons (4.5 and 13 Ml) of water per well, with large projects using up to 5 million US gallons (19 Ml). Additional water is used when wells are refractured; this may be done several times.

12.1   Hose materials

The traditional solution is to use metal pipes to transfer water but these are costly to deliver and assemble. Thermoplastic polyurethane (TPU) covered hoses and Nitrile rubber (NBR) covered hoses have a lot of advantages compared to metal pipes. It is easy to deliver, assemble and disassemble hoses and long sections of NBR- and TPU-covered hoses reduce the possibility of leakage in connecting parts. These large-diameter NBR- and TPU-covered hoses are called fracking hoses.

12.2   Hose selection

Fracking hoses are high-pressure, high-strength lay-flat hoses designed for pumping aggressive water around mine sites. Safety and efficiency are important factors when choosing products for these applications.

12.3   History

Fracking hoses are solutions derived from the fire hose industries. The fire hose manufacturers manufacture NBR or TPU covered hoses for fire fighting, water discharge, irrigation etc. The fracking industries borrowed the idea from the fire hose industries and use these hoses in fracking.

12.4   References

Chapter 13

The FracTracker Alliance

The FracTracker Alliance is a 501(c)(3) non-profit that shares maps, images, data, and analysis related to the oil and gas industry hoping that a better informed public will be able to make better informed decisions regarding the world's energy future.[1] FracTracker's information is focused in large part on unconventional extraction methods.[2] The FracTracker Alliance is based in the United States and has offices in Camp Hill, Pennsylvania; Pittsburgh, Pennsylvania; Ohio; New York; Berkeley, California; and West Virginia.[1]

FracTracker reportedly aims to provide non-partisan information, and has no official position on the practice of hydraulic fracturing.[2]

13.1 History

The FracTracker Alliance originated as FracTracker.org, a project of the Center for Healthy Environments and Communities at the University of Pittsburgh Graduate School of Public Health with the objective of crowd-sourcing data concerning unconventional gas extraction from the Marcellus Shale.[3] Between 2010 and early 2012, FracTracker was funded by grants from The Heinz Endowments and The William Penn Foundation.[3]

FracTracker.org's original director, Dr. Volz, left the University of Pittsburgh in April of 2011.[4] Soon after Dr. Volz left the University, FracTracker.org split off from the University as well (in early 2012), and formed a new non-profit named The FracTracker Alliance.[3] Many members of the original team from the University of Pittsburgh who had been working on FracTracker.org left the University to work for the new non-profit as (or shortly after) it was created.[3]

13.2 Current Initiatives

13.2.1 Mapping

FracMapper is the mapping component of FracTracker.org.[5] It offers a variety of maps detailing drilling-related activity. FracMapper's data comes from a variety of sources including state environmental agencies, news reports, freedom of information requests, user reports, collaborations with other groups, and information from other agencies. FracMapper makes its data available for download, and makes clear where the data came from - it includes a variety of metadata along with its data, including information about who created the original content, what is included in the dataset, when the dataset was taken, where the data features were located, and information about any changes from the original dataset.[6] FracTracker.org also offers regular in-person trainings about how to use FracMapper.[7]

FracTracker.org originally used a proprietary mapping system designed by Rhiza Labs in Pittsburgh,[8] but transitioned to a customized mapping platform based on Esri's ARCGis Online that allows users to download full datasets both through FracTracker and through Esri.[5]

13.3 References

[1] "About Us". FracTracker. Retrieved 6 February 2014.

[2] Pryts, Monica (3 April 2013). "FracMapper site lets users pinpoint local gas/oil drilling". *Sharon Herald*. Retrieved 6 February 2014.

[3] Kobell, Rona (1 February 2012). "Website collects, lets people track shale-drilling data in their area". *Bay Journal*. Retrieved 6 February 2014.

[4] Smit, Deb (17 April 2011). "What's next for Dan Volz?". *PopCityMedia*. Retrieved 6 February 2014.

[5] Malone, Sam. "Unveiling FracMapper, FracTracker's new mapping system!". Retrieved 6 February 2014.

[6] "Data Statement". FracTracker. Retrieved 7 February 2014.

[7] Pryts, Monica (12 April 2013). "FracMapper pinpoints local gas/oil drilling N". Retrieved 7 February 2014.

[8] "Concerned Groups Launch FracTracker.org to Assess Impacts of Marcellus Shale Drilling". 28 June 2010. Retrieved 6 February 2014.

13.4 See also

- GIS
- Crowdsourcing
- Marcellus Shale
- Hydraulic Fracturing
- Public Health
- Environmental Health

13.5 External Links

- FracTracker's website

Chapter 14

Fracturing Responsibility and Awareness of Chemicals Act

The **Fracturing Responsibility and Awareness of Chemicals Act** (H.R. 1084, S. 587, dubbed as the **FRAC Act**) is a legislative proposal in the United States Congress to define hydraulic fracturing as a federally regulated activity under the Safe Drinking Water Act. The proposed act would require the energy industry to disclose the chemical additives used in the hydraulic fracturing fluid. The gas industry opposes the legislation.[1]

The bill was introduced to both houses of the 111th United States Congress on June 9, 2009. The House bill was introduced by representatives Diana DeGette, D-Colo., Maurice Hinchey D-N.Y., and Jared Polis, D-Colo. The Senate version was introduced by senators Bob Casey, D-Pa., and Chuck Schumer, D-N.Y. The bill was re-introduced to both houses of the 112th United States Congress on March 15, 2011, by representative Diana DeGette and senator Bob Casey.

14.1   Background

The Environmental Protection Agency (EPA) blames the lack of information about the contents of hydraulic fracturing fluid on the 2005 Energy Policy Act because it exempts hydraulic fracturing from federal water laws.[2] The Act calls for the "chemical constituents (but not the proprietary chemical formulas) used in the fracturing process." Once these constituents are determined the information must be revealed to the public through the Internet. The FRAC Act states that in any case where a physician or the State finds that a medical emergency exists, and that the chemical formulas are needed to treat the ailing individual, the firm must disclose the chemical identity to the State or physician—even if that proprietary formula is a trade-secret chemical. Material Safety Data Sheets, required by OSHA under 29 CFR 1910.1200 are developed and made available to first responders and other emergency planning and response officials.

The drilling industry does not agree with this pending policy. They see it as "an additional layer of regulation that is unneeded and cumbersome."[3] The Independent Petroleum Association of America believes that states already sufficiently regulate hydraulic fracturing. Their research suggests that federal regulation could result in the addition of about $100,000 to each new natural gas well.[4] Energy in Depth, a lobbying group, says the new regulation would be an "unnecessary financial burden on a single small-business industry, American oil, and natural gas producers." This group also claims that the FRAC Act could result in half of the United States oil wells and one third of the gas wells being closed. Also, the bill could cause domestic gas production to drop by 245 billion cubic feet per year along with four billion dollars in lost revenue to the federal government.[5] The Environmental Protection Agency claims that the Safe Drinking Water Act is flexible in that it defers regulation of fracturing and drilling to the state. According to an industry-funded study, since most states currently have regulations on fracturing, they would most likely agree with the state's policy and there would not be much change.[4]

14.2 Current status

The 111th United States Congress adjourned on January 3, 2011, without taking any significant action on the FRAC Act. The FRAC Act was re-introduced in both houses of the 112th United States Congress. In the Senate, Sen. Bob Casey (D-PA) introduced S. 587 on March 15, 2011.[6] In the House, Rep. Diana DeGette (D-CO) introduced H.R. 1084 on March 24, 2011.[7]

Congress had not yet passed either of The FRAC Act bills[8][9] As of June 2013 the bill's status was "Died (Referred to Committee)."[10]

The FRAC Act was reintroduced as S. 1135 on Jun 11, 2013, and it only has a "9% chance of getting past [the] committee, [and a] 1% chance of being enacted."[11]

14.3 See also

- Hydraulic fracturing in the United States

- Environmental impact of hydraulic fracturing in the United States

- Proppants and fracking fluids

14.4 References

[1] O'Hehir, Andrew. "The FRAC Act under attack - Environment". Salon.com. Retrieved 2012-03-27.

[2] Environmental Effects of Hydraulic Fracturing, Hunter Valley Protection Alliance, 2008.

[3] Advanced Resources International. Potential Economic and Energy Supply Impacts of Proposals to Modify Federal Environmental Laws Applicable to the U.S. Oil and Gas Exploration and Production Industry. Prepared for U.S. Department of Energy. January, 2009.

[4] Lustgarten, Abraham. FRAC Act — Congress Introduces Twin Bills to Control Drilling and Protect Drinking Water. ProPublica. June 9, 2009.

[5] Tronche, John Laurent. "U.S. Representatives Unveil FRAC Act to close 'Halliburton Loophole.'" Fort Worth Business Press. June 9, 2009. www.fwbusiness.com.

[6] H.R. 1084: S. 587: FRAC Act

[7] H.R. 1084: Fracturing Responsibility and Awareness of Chemicals Act of 2011

[8] "H.R. 1084: Fracturing Responsibility and Awareness of Chemicals Act of 2011". GovTrack.us. Retrieved 9 March 2012.

[9] "S. 587: FRAC Act". GovTrack.us. Retrieved 9 March 2012.

[10] https://www.govtrack.us/congress/bills/112/s587

[11] https://www.govtrack.us/congress/bills/113/s1135

14.5 External links

- PDF of the House bill

- PDF of the Senate bill

Chapter 15

Hydraulic fracturing proppants

A **proppant** is a solid material, typically sand, treated sand or man-made ceramic materials, designed to keep an induced hydraulic fracture open, during or following a fracturing treatment. It is added to a *frac'ing fluid* which may vary in composition depending on the type of fracturing used, and can be gel, foam or slickwater–based. In addition, there may be unconventional frac'ing fluids. Fluids make tradeoffs in such material properties as viscosity, where more viscous fluids can carry more concentrated proppant; the energy or pressure demands to maintain a certain flux pump rate (flow velocity) that will conduct the proppant appropriately; pH, various rheological factors, among others. In addition, fluids may be used in low-volume well stimulation of high-permeability sandstone wells (20k to 80k gallons per well) to the high-volume operations such as shale gas and tight gas that use millions of gallons of water per well.

Conventional wisdom has often vacillated about the relative superiority of gel, foam and slickwater fluids with respect to each other, which is in turn related to proppant choice. For example, Zuber, Kuskraa and Sawyer (1988) found that gel-based fluids seemed to achieve the best results for coalbed methane operations,[1] but as of 2012, slickwater treatments are more popular.

Other than proppant, slickwater fracturing fluids are mostly water, generally 99% or more by volume, but gel-based fluids can see polymers and surfactants comprising as much as 7 vol% , ignoring other additives. Other common additives include hydrochloric acid (low pH can etch certain rocks, dissolving limestone for instance), friction reducers, guar gum, biocides, emulsion breakers, emulsifiers, 2-butoxyethanol, and radioactive tracer isotopes.

15.1   Proppant permeability and mesh size

Proppants used should be permeable or permittive to gas under high pressures; the interstitial space between particles should be sufficiently large, yet have the mechanical strength to withstand closure stresses to hold fractures open after the fracturing pressure is withdrawn. Large mesh proppants have greater permeability than small mesh proppants at low closure stresses, but will mechanically fail (i.e. get crushed) and produce very fine particulates ("fines") at high closure stresses such that smaller-mesh proppants overtake large-mesh proppants in permeability after a certain threshold stress.[2]

Though sand is a common proppant, untreated sand is prone to significant fines generation; fines generation is often measured in wt% of initial feed. A commercial newsletter from Momentive cites untreated sand fines production to be 23.9% compared with 8.2% for lightweight ceramic and 0.5% for their product.[3] One way to maintain an ideal mesh size (i.e. permeability) while having sufficient strength is to choose proppants of sufficient strength; sand might be coated with resin,to form CRCS (Curable Resin Coated Sand) or PRCS (Pre-Cured Resin Coated Sands). In certain situations a different proppant material might be chosen altogether—popular alternatives include ceramics and sintered bauxite.

Sand used for fracturing, USGS, 2012

15.2 Proppant weight and strength

Increased strength often comes at a cost of increased density, which in turn demands higher flow rates, viscosities or pressures during fracturing, which translates to increased fracturing costs, both environmentally and economically.[4] Lightweight proppants conversely are designed to be lighter than sand (~2.5 g/cm^3) and thus allow pumping at lower pressures or fluid velocities. Light proppants are less likely to settle. Porous materials can break the strength-density trend, or even afford greater gas permeability. Proppant geometry is also important; certain shapes or forms amplify stress on proppant particles making them especially vulnerable to crushing (a sharp discontinuity can classically allow infinite stresses in linear elastic materials).[5]

15.3 Proppant deposition and post-treatment behaviours

Proppant mesh size also affects fracture length: proppants can be "bridged out" if the fracture width decreases to less than twice the size of the diameter of the proppant.[2] As proppants are deposited in a fracture, proppants can resist further fluid flow or the flow of other proppants, inhibiting further growth of the fracture. In addition, closure stresses (once external fluid pressure is released) may cause proppants to reorganise or "squeeze out" proppants, even if no fines are generated, resulting in smaller effective width of the fracture and decreased permeability. Some companies try to cause weak bonding at rest between proppant particles in order to prevent such reorganisation. The modelling of fluid dynamics and rheology of fracturing fluid and its carried proppants is a subject of active research by the industry.

15.4   Proppant costs

Though good proppant choice positively impacts output rate and overall ultimate recovery of a well, commercial proppants are also constrained by cost. Transport costs from supplier to site form a significant component of the cost of proppants.

15.5   Other components of fracturing fluids

Other than proppant, slickwater fracturing fluids are mostly water, generally 99% or more by volume, but gel-based fluids can see polymers and surfactants comprising as much as 7 vol% , ignoring other additives.[6] Other common additives include hydrochloric acid (low pH can etch certain rocks, dissolving limestone for instance), friction reducers, guar gum,[7] biocides, emulsion breakers, emulsifiers, and 2-Butoxyethanol.

Radioactive tracer isotopes are sometimes included in the hydrofracturing fluid to determine the injection profile and location of fractures created by hydraulic fracturing.[8] Patents describe in detail how several tracers are typically used in the same well. Wells are hydraulically fractured in different stages.[9] Tracers with different half-lives are used for each stage.[9][10] Their half-lives range from 40.2 hours (lanthanum-140) to 5.27 years (cobalt-60).[11] Amounts per injection of radionuclide are listed in The US Nuclear Regulatory Commission (NRC) guidelines.[12] The NRC guidelines also list a wide range or radioactive materials in solid, liquid and gaseous forms that are used as field flood or enhanced oil and gas recovery study applications tracers used in single and multiple wells.[12]

In the US, except for diesel-based additive fracturing fluids, noted by the American Environmental Protection Agency to have a higher proportion of volatile organic compounds and carcinogenic BTEX, use of fracturing fluids in hydraulic fracturing operations was explicitly excluded from regulation under the American Clean Water Act in 2005, a legislative move that has since attracted controversy for being the product of special interests lobbying.

15.6   See also

- List of additives for hydraulic fracturing

15.7   References

[1] Mader, Detlef (1989). *Hydraulic proppant fracturing and gravel packing*. Amsterdam: Elsevier. ISBN 0-444-87352-X.

[2] "Physical Properties of Proppants". *CarboCeramics Topical Reference*. CarboCeramics. Retrieved 24 January 2012.

[3] "Critical Proppant Selection Factors". *Fracline*. Hexion.

[4] Rickards, Allan; et al. (May 2006). "High Strength, Ultralightweight Proppant Lends New Dimensions to Hydraulic Fracturing Applications". *SPE Production & Operations* **21** (2): 212–221.

[5] Guimaraes, M. S.; et al. (2007). "Aggregate production: Fines generation during rock crushing" (PDF). *Journal of Mineral Processing*.

[6] Hodge, Richard. "Crosslinked and Linear Gel Comparison" (PDF). *EPA HF Study Technical Workshop*. Environmental Protection Agency. Retrieved 8 February 2012.

[7] Ram Narayan (August 8, 2012). "From Food to Fracking: Guar Gum and International Regulation". *RegBlog*. University of Pennsylvania Law School. Retrieved 15 August 2012.

[8] Reis, John C. (1976). *Environmental Control in Petroleum Engineering*. Gulf Professional Publishers.

[9]   Scott III, George L. (3 June 1997) US Patent No. 5635712: Method for monitoring the hydraulic fracturing of a subterranean formation. US Patent Publications.

[10]   Scott III, George L. (15-Aug-1995) US Patent No. US5441110: System and method for monitoring fracture growth during hydraulic fracture treatment. US Patent Publications.

[11]   Gadeken, Larry L., Halliburton Company (08-Nov-1989). Radioactive well logging method.

[12]   Jack E. Whitten, Steven R. Courtemanche, Andrea R. Jones, Richard E. Penrod, and David B. Fogl (Division of Industrial and Medical Nuclear Safety, Office of Nuclear Material Safety and Safeguards (June 2000). "Consolidated Guidance About Materials Licenses: Program-Specific Guidance About Well Logging, Tracer, and Field Flood Study Licenses (NUREG-1556, Volume 14)". US Nuclear Regulatory Commission. Retrieved 19 April 2012. labeled Frac Sand...Sc-46, Br-82, Ag-110m, Sb-124, Ir-192

Chapter 16

Hydro-slotted perforation

Hydro-slotting perforation technology is the process of opening the productive formation through the casing and cement sheath to produce the oil or gas product flow (intensification, stimulation). The process has been used for industrial drilling since 1980, and involves the use of an underground hydraulic slotting engine (tool, equipment). The technology helps to minimize compressive stress following drilling in the well-bore zone (which reduces the permeability in the zone).

16.1   Overview

Since ancient times, when there were the first coal mines, it was observed, that increasing the depth of the development the coal tunnel, under the action of overburden pressure, surrounding rocks become harder and little-permeable. To solve this problem they developed a cavern of a certain form in the rock. More modern mining geo-mechanics explain the reason for the occurrence of this effect in relation to drilling wells. During any drilling process in the well there is formed the annular compressive stress conditions around the wellbore zone. The deeper the well, the more overburden pressure, which means the greater the annular compressive stress conditions. On the rocks lying at depths of 3–5 km the compressive stresses may reach up-to 75–125 MPa. In the near-well zone, as a result of concentration these stresses increase and sometimes become equal to double 150–250 MPa. If the tectonic stresses is several times higher than stresses from the weight of rocks, the stresses in the near-well zone may be even greater.

Under the action of stress conditions and high overburden pressure occurs a significant reduction in permeability in the near wellbore zone, in some cases close to zero. Oil or gas flow can not penetrate to the well. Traditional methods of opening the productive layer formation (cumulative, jet perforation, sand jet perforation, abrasive jetting perforation and other similar methods) did not consider this complicated situation in the near-well zone and therefore was not the effective. Porous and fractured formations are subjected to compression, that deforms the rock mass and reduces its permeability. The greater the depth, the stronger the effect can be.

Hydro-slotting perforation is quite different from jet (hydro-jetting or sand-blast) perforation. The energy of working fluid, consisting from water (layer water) and sand (abrasive quartz sand) pressure in the hydraulic engine is divided into two components: five percent of energy goes to the creation of smooth uniform rectilinear motion of the working rod with the perforator and nozzles (between two and six nozzles) without participation in the process the multimeter tubing or coil-tubing. Ninety-five percent of energy is goes to the cutting of continued and geometrically correct deep slots (up to five feet deep and between three to five slots at the same time). Slot length is equal to the length of the working engine shaft, usually 1.64 feet (0.00050 km).

The hydro-slotting perforation process does not deform the casing, does not create cracks in the cement, and does not clog-up the borders in the formation.

The geometry and depth of the slots creates the conditions for occurrence of the effect of unloading the circular stress conditions in the near wellbore zone (from 50 to 100 percent) and accordingly the increase of permeability (up to 30 to 50 percent) in this zone. In addition to this it forms a large area of the penetration (31.5 square feet (2.93 m^2) area for one cut with two nozzles only), that provides a very good hydrodynamic connection of the productive layer with the well.

The cutting speed may be corrected with the temperature in the borehole, temperature of the working fluid, concentration, flow and pressure. (these components are enough to completely control the depth and length of the cut and thus forming the slots), to instantly cut through the steel casing, through the cement to delve into the productive formation and keep the jets in this state while moving along the borehole, keeping the same depth of cut. At the end of the cutting continuous slot process the engine is set up to the initial position and ready for the next cutting interval. The process of hydro-slotting perforation and the depth of cut is controlled by the working fluid supply, pressure and concentration. The equipment can be operated without lifting on the surface for 11–15 hours.

Hydro-slotting perforation is the ecologically safe, environmentally friendly and effective affordable method for intensifying the operation in oil, gas, injection and hydro-geological wells. Now this method is widely used in Azerbaijan, Brazil, China, Eastern and Western Siberia, Jordan, Kazakhstan, Komi Republic, North Caucasus, Russia, Udmurtia, Ukraine, Urals, Uzbekistan and Yemen. The first mention regarding the hydro-slotting perforation in America, was in 1987 at the oil and gas conference in Texas. The first use of hydro-slotting perforation in the United States dates back to 1996, when together with Shell E & P Technology Company, discovered two wells (Abrasive Hydro jet Technology in Albert Load, Michigan). After that the hydro-slotting perforation was highly appreciated by the Department of Geophysics at Stanford University and by Division of Shell Exploration and Production by Shell E&P Technology Company. Hydro-slotting perforation was used in California, Kansas, Michigan, Montana, Nebraska, New York, Pennsylvania, Texas and Wyoming states. In Canada it has been successfully applied in Saskatchewan.

16.2 General concepts

For opening of any productive layer it is necessary to open the casing, cement sheath and productive layer formation. Geophysics and mining geo-mechanics dictated the next requirements:

- Zone of cement sheath should be opened completely and not have cracks (to prevent possible overflows of water);

- Productive layer formation should be opened to maximum and on the maximum depth. At the same time productive formation should not have clogging, plugging, grouting, occlusive and cinder borders to produce excellent hydrodynamic connections of the productive layer with the well. Encompassing unloading the circular stress conditions around the wellbore, formed as a result of drilling, and increasing the permeability (50–100 percent) in the near wellbore zone (as a consequence of the first)

In the early 1970s, the Ministry of Geology of the USSR placed the Government order to scientific research institutions of the Country for the solution of the annular stress conditions and increase the permeability problem in the drilling wells. It was necessary to create the technology of opening the productive layer formation taking into account of uploading the annular stress conditions and increase the permeability in the near wellbore zone. The work to study this problem were assigned to the Institute of Oceanology and VNIMI (St. Petersburg, Russia). During the study there was done hundreds of experiments and mathematical models. It was determined, that if creating a geometrically correct, extended slot, directed along the wellbore and perpendicular to it on the distance from around 0.7 inches (18 mm) to 3.5 feet (1.1 m), in the zone of around the wellbore, there occurs the unloading of the annular compressive stress conditions from 50 to 100 percent, that are redirected to the far plane of the surface of formed slot, parallel to the wellbore surface. At the same time the permeability in this zone increased 30 to 50 percent. The holes after cumulative, jet perforation, sand jet perforation, abrasive jetting perforation and other similar methods, do not give the effect. The spot perforation did not create a slot in the casing and did not reach the required (unloading effect) depth, because the reverse jet interfered with direct jet and the maximum depth of the hole could not exceed 0.65 feet. When perforation with the movement occurs, the direct jet does not intersect with the reverse jet and depth of cutting can be much more (up to five feet) which is known as the excavation effect. Later it was proven mathematically.

It was necessary to create a device, that could make the continuation, along the borehole, slots in the casing, cement and go further into the productive formation. The tests with the movement of the multimeter tubing were not been successful, showing it was impossible to create geometrically correct extended slots with moving tubing. It was necessary to create an apparatus, that created a movement of cutting jets by itself, independently from tubing and located on the end of the tubing, directly in the leveled area. independent movement of the cutting jets could only be done mechanically, electrically or hydraulically. After another six months of research and testing it was decided to use mechanics and hydraulics as the base.

The first prototype of hydro-slotting perforation device was created in 1972. The technology of hydro-slotting perforation was never sold to anyone. The hydro-slotting perforation technology was transformed into the category of performance techniques (as the technique of conducting the drilling, cumulative perforation, hydraulic fracturing, logging, pumping and so on).

The finalization of the device (prototype) in the end of 1972 was tasked to the special laboratory of the Research Institute of Oceanology of PSU "Sevmorgeo".[1] From the beginning of the work for the revision the existing device was carried out in two directions: hydro-slotting perforation and hydro-mechanical slotting perforation. The second variant differs from the first in that at the beginning the opening of the casing is produced with a circular saw, and then the rock eroded by working fluid (water and sand) jets. The works were done over three years. The work for improvements of the hydro-mechanical slotting perforation were terminated in the result of their further inexpedient. Firstly, it was not necessary to divide the process into two operations: cutting the casing with the circular saw and a further jet-slotting perforation, because the cutting of casing with jet-slotting perforation takes place in a matter of seconds. Second, the mechanism of the circular saw takes up a lot of space in the housing unit, it was impossible to use the energy of working fluid to full power for getting deep slots, the slots get small and not deep (not enough for occurs the unloading of the annular compressive stress conditions and increase the permeability in the near wellbore zone). The further project was focused for finishing the hydro-slotting perforation device only.

In 1975, the scientific research laboratory of the Research Institute of Oceanology of PSU Sevmorgeo[1] completed the project to improve the prototype of hydro-slotting perforation tool and this tool has been able to operate independently of the tubing movement. The equipment was 16 feet (4.9 m) long, 4.02 inches (10.2 cm) OD, weight 300 pounds (140 kg) and stroke length of 0.5 feet (0.15 m) only, and it worked on the following principle: the energy of working fluid pressure was divided into two components. Part of energy was used for the motion creation for the working rod with the perforator and nozzles; the other part of the energy was use for the cutting process (creating the continued slots along the wellbore through the casing and cement into the productive formation). The form and depth of the slots allowed the device to perform its main task, unloading the annular stress conditions and increase the permeability. The first practical tests in the wells were successfully made at the end of 1975 on "Archeda" field (Volgograd, Russia).

16.2.1 Benefits

Ability to increase area of development

- Very deep penetration from three to six feet

- Vertical permeability

- Porosity increases four to five times

- Permeability increases 15 times

- Drainage volume increases six times

Ability to access reserves which are otherwise inaccessible

- In reservoirs located in close proximity to water, gas, and oil contacts

- In weakly permeable, tightly-cemented reservoirs

- In missed layers, or in layers covered by two or more columns

Gentle approach with the ability to repair well-bore damage

- In carbonates, dislodges clay particles and fines

- In sandstones, reduces sand mobility problems

- In deep gas sands, relieves overpressure damage from mud weight systems

- Does not crack casing or cement

- Maintains hydraulic integrity with no detonation impacts

- Redistributes stresses away from the near-well-bore zone

16.3   Development

During the period from the date of the first prototype of hydro-slotting perforation tool to present day, the type and techno-logical characteristics of the equipment was significantly improved. The modern underground hydro-slotting equipment represents the devices, capable to instantly cut through the steel casing, through the cement to delve into the productive formation and keep the jets in this state while moving along the borehole, keeping the same depth of cut. Hydro-slotting equipment made of special high-strength materials, 12 feet (3.7 m) long, 3.5 feet (110 cm) OD, weight 180 pounds (82 kg), cutting speed from the point of perforation to 0.7 inches (1.8 cm) per minute, working stroke length 1.65 feet (0.50 m) (4.92 feet (1.50 m) x 1.64 feet (0.50 m) x 1.97 inches (5.0 cm) each slot), depth of slots five feet, continued and geo-metrically correct slots, opening area 63 square feet (5.9 m^2) per cut with four nozzles, can apply streamlined perforators between two and six nozzles, unloading the annular stress conditions in the near wellbore zone 50 to 100 percent, and increase the permeability 30 to 50 percent. The continuous time without lifting to the surface is 11–15 hours (nozzles lifespan ~ 15 hours, perforator ~ seven wells, hydraulic engine ~ 40 wells).

Without lifting to the surface with the hydro-slotting tool can also:

- cut on the previous perforation (cumulative perforation)

- colmatation treatment

- cut the thin-interbedded layers

- mini hydraulic fracture stimulation

- create the continuous slot

- cut the shale

- accurate cut near the water reservoir or opposite in the injection wells

- bypass the water layers

- bypass the casing collars

- cut a few casings

- chemical treatment

- sealing, direct and reverse flushing

- tubing pressure testing

- cut the casing at abandonment

The hydro-slotting perforation process does not deform the casing, does not create cracks in the cement and does not clog up the borders in the formation. The process of hydro-slotting perforation is controlled. The cutting speed and depth of cutting may be corrected with the temperature in the borehole, temperature of the working fluid, concentration, flow and pressure. At the end of the cutting process of a continuous slot the engine is set up to the initial position and ready for the next cutting interval. Hydro-slotting perforation sets the perfect geometry for the subsequent fracturing. Hydro-slotting perforation can be applied in any formation: shale, carbonates, sandstone and so on.

Further improvement of the equipment for hydro-slotting perforation must follow the scientific and technical progress in this technology, not on the way of mindless increase of the holes in the hydro jets pipe. It is necessary to make the underground hydraulic engine for horizontal wells, which must be sealed to prevent the ingress of sand and mud inside and maintain the centerline position relative to the wellbore. It is necessary to make a self-orientation perforator (a particularly important issue of orientation in horizontal wells). For the orientation of the tool it is necessary there is communication with the tool (preferably two-sided) and surface of the well. Taking into account the specific conditions of hydro-slotting perforation process, signaling from the tool and back possibly using ultrasound only. Then the cutting process can be fully controlled from the surface, and it will be possible to change the speed and depth of cutting the slots regardless of the temperature inside the well.

16.4   Patents

Over the years this method has not undergone much change, but there are many patents on the method of hydro-slotting perforation. With the development of technological progress there has been continuously improved and refined equipment, but patents, regarding the hydro-slotting equipment in full is not so much, there are a few patents on parts.

- United States patents for complete hydro-slotting perforation equipment: US 8240369 B1, US 31,084

- Similar United States patents: US3130786, US4227582, US5337825, US6651741, US7073587, US7140429, US7568525, US20070187086, US20090101414, USRE21085, 166/55.2, 166/298, and E21B43/114

- United States patent for method of hydro-slotting perforation: US 20130105163 A1

- Similar United States patents: US3130786, US4047569, US4134453, US5445220, US6564868, US7568525, and US20050269100

16.5   References

[1] *Oil and Gas Industry Magazine*, January 2008

Chapter 17

Mohamed Yousef Soliman

Mohamed Yousef Soliman[6] is a professor and the former chairperson of the department of Petroleum Engineering at Texas Tech University. After working for Halliburton for 32 years, he joined Texas Tech in January 2011. He obtained his bachelor's degree in petroleum engineering from Cairo University in 1971. Having completed his bachelor's degree he came to the United States to continue higher education. He received his masters and doctorate degrees, both in Petroleum engineering, from Stanford University in 1975 and 1978 His M. S. Thesis was "Rheological Properties of Emulsion Flowing Through Capillary Tubes Under Turbulent Conditions,"; his Ph. D thesis, "Numerical Modeling of Thermal Recovery Processes."

He h is the author or co-author of over 180 technical papers, almost all of which are in the field of Petroleum engineering and Oil industry. Additionally he is credited with 20 inventions.

17.1 Biography

Soliman is a specialist in r hydraulic fracturing and production engineering. He holds 21 patents on Hydraulic fracturin operations and analysis, testing and conformance applications,[7] and is an author or co-author of over 170 technical papers and articles in areas of fracturing, reservoir engineering, well test analysis, conformance, and numerical simulation.

He has written chapters in *World Oil's Handbook of Horizontal Drilling and Completion Technology*, the text *Well Construction*, and the SPE monograph *Well Test Analysis of Hydraulically Fractured Wells*. He has authored several books for internal use at Halliburton, including Stimulation and Reservoir Engineering Aspects of Horizontal Wells, Well Test Analysis, Hydraulic Fracturing, and chapters in Conformance, Stimulation, and FracPac.[8] He is a distinguished member of the Society of Petroleum Engineers.

17.2 Most recent peer-reviewed papers

1. Fahd Siddiqui, Mohamed Y. Soliman, Waylon House, "A new methodology for analyzing non-Newtonian fluid flow tests" *Journal of Petroleum Science and Engineering*, Volume 124, December 2014, Pages 173-179, ISSN 0920-4105 [9]

2. M. Y. Soliman, Johan Daal, and Loyd East. " Fracturing Unconventional Formations to Enhance Productivity." *Journal of Natural Gas Science and Engineering* Invitational paper. 8 (2012) 52-67.

3. M. Y. Soliman, and C. S. Kabir, 2012. "Testing unconventional formations." *J. of Petroleum Science & Engineering* Invitational paper. 92-93 (2012) 102-109.

4. Leopoldo Sierra, Loyd East, M.Y. Soliman, and David Kulakofsky. 2011New Completion Methodology To Improve Oil Recovery and Minimize Water Intrusion in Reservoirs Subject to Water Injection, SPE Journal, September 2011.

5. Loyd East Jr., M.Y. Soliman, and Jody Augustine. 2011. Methods for Enhancing Far- Field Complexity in Fracturing Operations. *SPE Production Operations*, August 2011.

6. M.Y. Soliman, Carlos Miranda, Hong (Max) Wang, and Kim Thornton, Investigation of Effect of Fracturing Fluid on After-Closure Analysis in Gas Reservoirs. Production & Operations SPE Journal, May 2011

17.3   References

[1] "Bob L. Herd Biography".

[2] "Texas Tech University Petroleum Engineering Academy Members".

[3] "Bob L. Herd Department of Petroleum Engineering, Texas Tech University".

[4] "Lioyd Heinze Homepage".

[5] "Marshall Watson Homepage".

[6] "Mohamed Y. Soliman Personal homepage".

[7] B1 US US6283210 B1, Mohamed Yousef Soliman; Steve Rester & Michael H. Johnson et al., "Proactive conformance for oil or gas wells", published Sep 4, 2001, assigned to Halliburton Energy Services, Inc.

[8] "Mohamed Y. Soliman Curriculum Vitae" (PDF).

[9] http://dx.doi.org/10.1016/j.petrol.2014.10.007.

Chapter 18

Oil and Gas Commission

The **Oil and Gas Commission** (OGC) is a Crown Corporation of the province of British Columbia, Canada, established in 1998. Its mandate is to regulate oil and gas activities and pipelines in British Columbia. Their mandate does not extend to regulating consumer gas prices at the pump.

18.1 Overview

The Oil and Gas Commission (OGC) was created and defined under the 1998 *Oil and Gas Commission Act* by and for the Canadian province of British Columbia.[1]

The OGC is a crown corporation acting as "an agent of the [provincial] government", where the Minister of Finance is its fiscal agent. It is headed by 3 directors. The deputy minister is a director and the chair of the OGC, and the Lieutenant Governor in Council may appoint 2 directors, for a term not longer than 5 years, one of whom is the commissioner and vice chair of the commission.[1]

18.1.1 Purpose

The OGCs purposes are to

(a) regulate fossil fuel activities, i.e. the petroleum industry, which includes but is not limited to petroleum licensing, regulating hydrocarbon exploration, oil and gas well drilling, shale oil extraction, natural gas processing, oil pipeline in British Columbia in a manner that

- "provides for the sound development of the oil and gas sector, by fostering a healthy environment, a sound economy and social well being,"

- "conserves oil and gas resources in British Columbia",

- "ensures safe and efficient practices, and"

- "assists owners of oil and gas resources to participate equitably in the production of shared pools of oil and gas,"

(b)" provide for effective and efficient processes for the review of applications related to oil and gas activities or pipelines, and to ensure that applications that are approved are in the public interest having regard to environmental, economic and social effects",

(c) "encourage the participation of First Nations and aboriginal peoples in processes affecting them",

(d) "participate in planning processes", and

(e) "undertake programs of education and communication in order to advance safe and efficient practices and the other purposes of the commission."[1]

18.1.2   Tools

The OGC issues various authorizations under the *Petroleum and Natural Gas Act* and the *Pipeline Act,* including a "general development permit" which is "an approval in principle for oil and gas activities and pipelines in an area of British Columbia".[1]

The OGC is a single window regulator[2] It handles applications and at the same checks the compliance and enforces its regulations with penalties for violations.

The OGC has offices in four cities: Fort St. John, Fort Nelson, Kelowna and Victoria.

18.1.3   Compliance and enforcement information

In 2010-11, the OGC "issued 15 penalty tickets with fines of $575 (the maximum allowed for tickets) or less, which included unlawful water withdrawals and failure to promptly report a spill. Court prosecutions included a $20,000 fine for a Water Act stream violation, $10,575 for another stream violation and $250,0000 for a sour gas release. [...] The commission would not release the names of the companies convicted".[2] Per the OGC, in 2012, of "more than 800 deficiencies, 80 resulted in charges, largely under the provincial Water Act for the non-reporting of water volumes and a smaller portion under the provincial Environment Management Act. Another 13 resulted in orders under the provincial Oil and Gas Activities Act, 22 in warnings, 76 in letters requiring action and three in referrals to other agencies".[2] Paul Jeakins, OGC commissioner and CEO, has publicly acknowledged that OGC inspection and enforcement reports are "a bit of a gap".[2]

18.2   Lawsuits

In November 2013, Ecojustice, the Sierra Club and the Wilderness Committee filed a lawsuit against the OGC and Encana about Encana's water use from lakes and rivers for its hydraulic fracturing for shale gas, "granted by repeated short-term water permits, a violation of the provincial water act".[3] In 2012, the OGC had granted Encana access to 20.4 million cubic metres of surface water, 7 million of which were for fracking and 54% of that were through short-term approvals.[4] In October 2014 the Supreme Court of British Columbia found no violation and dismissed the case.[4]

18.3   Criticism

The agency has been criticized to be "too industry-friendly", to have "vague regulations" and to issue non transparent fracking violation reports, for example by not naming convicted companies. The B.C. Ministry of Environment and other B.C. Crown corporations of B.C. like WorkSafeBC have reported company names and details of those penalties for years. OGC reports prior to 2011 were available on the OGC website, but for 2012 they had to be requested. [2]

18.4   See also

- History of the petroleum industry in Canada (frontier exploration and development)

- Greater Sierra (oil field) in B.C.

- List of Canadian natural gas pipelines

- List of Canadian oil pipelines

- List of Canadian pipeline accidents

- List of oil spills

- List of largest oil and gas companies by revenue

- Oil and gas law in the United States

- Petroleum fiscal regime

- Pipeline and Hazardous Materials Safety Administration- US counterpart regarding pipelines

18.5   References

[1] "Oil and Gas Commission Act". Queen's Printer, Victoria, British Columbia, Canada. Retrieved 4 February 2015.

[2] Gordon Hoekstra (18 February 2013). "B.C. Oil and Gas Commission lacks 'transparency' on fracking violations". *Vancouver Sun* (Postmedia Network Inc). Retrieved 4 February 2015.

[3] Dene Moore (13 November 2013). "Fracking Lawsuit Targets EnCana, B.C. Oil And Gas Commission". *Canadian Press* (The Huffington Post). Retrieved 1 August 2014.

[4] Dene Moore (16 October 2014). "B.C.'s Supreme Court dismisses Water Act challenge to fracking operations". *The Globe and Mail* (Phillip Crawley). Retrieved 4 February 2015.

18.6   External links

- Oil and Gas Commission - official site

Chapter 19

Promised Land (2012 film)

Promised Land is a 2012 American drama film directed by Gus Van Sant and starring Matt Damon, John Krasinski, Frances McDormand, and Hal Holbrook. The screenplay is written by Damon and Krasinski based on a story by Dave Eggers. *Promised Land* follows two corporate salespeople who visit a rural town in an attempt to buy drilling rights from the local residents.

Damon was originally attached to direct the film, but he was replaced by Van Sant. Filming took place mainly in Pittsburgh from early to mid-2012. During filming and afterward, the film's highlighting of the resource extraction process hydraulic fracturing, known as "fracking," emerged as a topic of debate.

The film had a limited release in the United States on December 28, 2012 and followed with a nationwide expansion on January 4, 2013. The film had its international premiere and received Special Mention Award at the 63rd Berlin International Film Festival in February 2013.

19.1 Plot

Steve Butler has caught the eyes of top management at his employer, Global Crosspower Solutions, an energy company that specializes in obtaining natural gas trapped underground through a process known as fracking. Butler has an excellent track record for quickly and cheaply persuading land owners to sign mineral rights leases that grant drilling rights over to his employer. Butler and his partner Sue Thomason arrive in an economically struggling Pennsylvania farming town whose citizens are proud of having family farms passed from one generation to the next. Coming from a town and a life very similar to that of the people he is now determined to win over on behalf of Global, Butler tells the story of how his own town died after the local Caterpillar assembly plant closed. The idea of a town surviving solely on family farms being passed down through generations as a viable economy is one that he can no longer accept. He claims to be offering the town its last chance. Butler spends some pleasant after-hours time with Alice, a teacher he meets in a bar.

The town decides to put Global's offer up to a community vote and seems willing to vote in Global's favor until a local high school science teacher, who happened to be a successful engineer in his working life, raises the question of the safety of fracking during a town meeting. Butler and Thomason's sales pitch is further challenged when Dustin Noble, an unknown environmental advocate, starts a grassroots campaign against Global, motivated by a tale of his family losing its dairy farm after the herd died as a result of Global's industry-standard fracking process.

Butler begins to meet a great deal of resistance in town. Noble seems to be winning over nearly everyone, including Alice. One night Butler receives a package from Global that includes an enlarged copy of a photograph of dead cattle on a field that Noble said came from his family's Nebraska farm. The enlargement shows that the object thought to be a silo is in fact a lighthouse and that Noble has been practising deception.

Butler tells the town's mayor and then visits Alice, trying to prove that "I'm not the bad guy." He returns to the hotel to find Noble is loading his truck and leaving town. Noble accidentally reveals that he knows the picture with the lighthouse was taken in Lafayette, Louisiana. The film makes no reference to the fact that Lafayette is a landlocked parish with no coastline. Butler realizes that the only way Noble could know this is if he were also employed by Global. Noble's job had been to discredit the environmental movement. He arranged for Butler to receive the "confidential" photos and engineered the entire public relations effort. Noble wishes Butler good luck back at the company's headquarters in New York.

At a town meeting the next day, the citizens are prepared to vote on Global's efforts to buy gas rights to their property. Butler tells how the barn in the picture reminds him of his grandfather's barn. He reveals that Noble had manipulated them and that Noble actually is employed by Global. Butler leaves the meeting to find Thomason on the phone with Global. She tells him that he's fired and that she is leaving for New York. Butler walks to Alice's home and she welcomes him in.

19.2 Cast

- Matt Damon as Steve Butler

- John Krasinski as Dustin Noble

- Frances McDormand as Sue Thomason

- Rosemarie DeWitt as Alice

- Scoot McNairy as Jeff Dennon

- Titus Welliver as Rob

- Terry Kinney as David Churchill

- Hal Holbrook as Frank Yates

19.3 Production

Promised Land is directed by Gus Van Sant based on a screenplay by Matt Damon and John Krasinski, who are film producers along with Chris Moore. The screenplay was based on a story by Dave Eggers. Krasinski came up with the film's premise and developed the idea with Eggers. They pitched the idea to Damon, suggesting that both Damon and Krasinski would write and star in the film. The project was set up at Warner Bros. with Damon attached as director in October 2011, in what would have been his directorial debut. Filming was scheduled to begin in early 2012.[1]

In January 2012, Damon stepped down as director due to scheduling conflicts but remained involved with the project.[2] Damon contacted Gus Van Sant, who directed him in the 1997 film *Good Will Hunting*, and Van Sant joined the project as director.[3] The project was in turnaround at Warner Bros., and by February, Focus Features and Participant Media acquired rights to produce the film. The title was announced to be *Promised Land*.[4] With a production budget of $15 million,[5] filming began in Pennsylvania in late April 2012.[6] The Commonwealth of Pennsylvania provided the production company $4 million in tax credits since filming would provide jobs and revenue.[7] Over eighty percent of the crew were hired out of Pittsburgh. Filming mostly took place in Avonmore, Pennsylvania, which was the main setting for the film's rural town of McKinley. Additional filming locations for the town were locations in Armstrong County including Apollo, Worthington, and Slate Lick. Other filming locations in Pennsylvania were Alexandria, Delmont, Export, and West Mifflin. Filming also took place at the Grand Concourse at Station Square in Pittsburgh. Several hundred extras were hired for the film, and filming lasted for 30 days.[8]

The movie was financed by Image Productions, a company owned by the Government of Abu Dhabi.

The film score was composed by Danny Elfman.[9] Three songs by The Milk Carton Kids including Snake Eyes, The Ash & Clay and Jewel of June were also written for the film.[10]

19.4 Fracking

Main article: Hydraulic fracturing

Promised Land was criticized by the energy industry for its portrayal of the resource extraction process hydraulic fracturing, colloquially known as "fracking".[11] The portrayal was first reported in April 2012 by filmmakers raising funds for the pro-fracking documentary *FrackNation*. They said, "*Promised Land* will increase unfounded concerns about fracking."[12] Phelim McAleer, the director of *FrackNation*, said Dimock, Pennsylvania was the likely inspiration for *Promised Land*. McAleer said despite Dimock families' claims that fracking activity contaminated their water, the state and EPA's scientists did not find anything wrong.[13] In September 2012, CNBC reported that a group of residents from Armstrong County, Pennsylvania were protesting the film and formed a Facebook group. The group said, "They filmed this movie in our backyard. They told us it would be fair to drilling. It's not. We're p*ssed [sic]."[14] Mike Knapp, one of the organizers of the Facebook group said, "One of the things that really aggravates me, is that they seem to have a very condescending view" of farmers as portrayed in the film.[7]

Krasinski, who co-wrote the screenplay and stars in *Promised Land*, said the film's original premise involved wind power. Krasinski said wind power was replaced by fracking as a more relevant backdrop based on news coverage in recent years.[8] *The Huffington Post* reported, "The procedure has caused concern due in part to the chemicals injected into the wells for drilling, which may taint nearby drinking water." It said Damon had posted in 2010 a YouTube video to promote the Working Families Party, which works "to prevent risky natural gas drilling".[15] *Politico* said *Promised Land* reflected a trend about fracking since the release of the 2010 documentary film *Gasland*, which was nominated for an Academy Award for Best Documentary.[12]

Leading up to the film's release, a spokesperson for Independent Petroleum Association of America said, "We have to address the concerns that are laid out in these types of films." The industry planned to send scientific studies to film critics, to distribute leaflets to film audiences, and to use social media like Facebook and Twitter as a response to the film.[11][16] Where the industry launched "direct attacks" at *Gasland*, it instead sought to portray *Promised Land* as "derivative, condescending and clichéd". In Pennsylvania, the industry group Marcellus Shale Coalition bought a 16-second onscreen ad to be shown at 75 percent of theaters in the state at the same time *Promised Land* was released.[7]

James Schamus, chief executive of the film's distributor Focus Features said, "We've been surprised at the emergence of what looks like a concerted campaign targeting the film even before anyone's seen it."[11] As the film was released, he said, "Fracking is a great premise for real drama. It represents Americans deeply conflicted about how to deal with these issues." He compared the industry's stealth campaign against the film to the one depicted within the film.[7]

19.5 Financing

The Heritage Foundation, a conservative think tank, reported that *Promised Land* was financed in part by Image Nation Abu Dhabi, a subsidiary of Abu Dhabi Media, which is wholly owned by the United Arab Emirates. The foundation said that the UAE, as a member of the Organization of Petroleum Exporting Countries (OPEC), has "a direct financial interest... in slowing the development of America's natural gas industry" and suggested that its financing of the film "may have an impact on the public's view of the [fracking] practice".[17][18] Image Nation said it provided financing to the film as part of an ongoing partnership with Participant Media, "regardless of genre or subject matter".[11][16]

19.6 Release

19.6.1 Theatrical run

Promised Land had a limited release on December 28, 2012, making it eligible for the 85th Academy Awards, but failed to win any.[19] The film was released in 25 theaters and grossed an estimated $53,000 on its first day, a "sobering" average of $2,120.[20] For the opening weekend, *Promised Land* grossed an estimated $190,000. Box Office Mojo reported before the film's wide release the following week, "It's unlikely that it will be able to pull many people away from the various other appealing options in theaters right now."[21] *Promised Land* expanded to 1,676 theaters on January 4, 2013. It grossed $4.3 million over the weekend, which the *Los Angeles Times* judged as "a bad start" even with its $15 million budget. According to CinemaScore, audiences gave the film a "B" grade. The *Times* said the grade and "middling reviews" indicated the film was unlikely to be a success.[22] By the end of its theatrical run, the film grossed $8.1 million, failing to make back its budget of $15 million.[23]

The film had its international premiere at the 63rd Berlin International Film Festival in February 2013[24] where Gus Van Sant won a Special Mention.[25]

19.6.2 Critical reception

Promised Land received mixed reviews from critics. *The Los Angeles Times* reported that most critics felt that the film did not reach its full potential.[26] On Rotten Tomatoes the film has a rating of 51%, based on 145 reviews, with an average rating of 5.8/10. The site's critical consensus reads, "The earnest and well-intentioned *Promised Land* sports a likable cast, but it also suffers from oversimplified characterizations and a frustrating final act."[27] Metacritic gave the film a score of 55 out of 100, based on 32 critics, indicating "mixed or average reviews".[28]

New York Times film critic A.O. Scott praised *Promised Land* as a film that "works" mainly "by putting character ahead of story" and by "inviting the actors to be warm, funny and prickly".[29] Liam Lacey of *The Globe and Mail* is critical of the film: "Apart from its warm, gentle tone, much about Promised Land simply isn't good, especially the inconsistencies in the screenplay. After the mood-setting first half, things start to unravel."[30]

19.7 Accolades

19.8 References

[1] Kit, Borys (October 19, 2011). "Matt Damon to Direct Warner Bros. Drama He Co-Wrote With John Krasinski". *The Hollywood Reporter*.

[2] Kit, Borys (January 5, 2012). "Matt Damon Won't Direct Movie He Co-Wrote with John Krasinski". *The Hollywood Reporter*.

[3] Kit, Borys (January 6, 2012). "Gus Van Sant Taking Matt Damon's Place on Dave Eggers Project". *The Hollywood Reporter*.

[4] Weinstein, Joshua L. (February 1, 2012). "Focus Features comes aboard Damon, Krasinski film". Reuters.

[5] Jagernauth, Kevin (February 1, 2012). "Gus Van Sant/Matt Damon's 'Promised Land' Goes To Focus & Participant". *The Playlist* (indieWire). Retrieved October 2, 2012.

[6] "Production Begins on Gus Van Sant's *Promised Land*". ComingSoon.net. April 24, 2012. Retrieved October 2, 2012.

[7] Drajem, Mark (January 4, 2013). "Fracker Ad Clashes on Screen With Damon's 'Promised Land'". *Bloomberg Businessweek*.

[8] Vancheri, Barbara (June 15, 2012). "John Krasinski and Gus Van Sant totally immersed in film shot in Pittsburgh". *Pittsburgh Post-Gazette*.

[9] Martens, Todd. "Journalist". LA Times. Retrieved 12/6/2012. Check date values in: |access-date= (help)

[10] Burlingame, Jon (2012-12-12). "Song: In Their Own Words". *Variety*.

[11] Gilbert, Daniel (October 7, 2012). "Matt Damon Fracking Film Lights Up Petroleum Lobby". *The Wall Street Journal*. Retrieved October 9, 2012.

[12] Buford, Talia; Martinson, Erica (April 5, 2012). "Matt Damon to star in 'The Promised Land' anti-fracking movie". *Politico*.

[13] McAleer, Phelim (September 25, 2012). "For his next escape". *New York Post*. Retrieved December 27, 2012.

[14] Carney, John (September 28, 2012). "Matt Damon's Anti-Fracking Film Backed by OPEC Member". *NetNet* (CNBC).

[15] "'Promised Land': Matt Damon's Fracking Film To Highlight Controversial Drilling Process". *The Huffington Post*. April 6, 2012. Retrieved October 2, 2012.

[16] Szalai, George (October 8, 2012). "Energy Industry Targets Upcoming Matt Damon Film 'Promised Land'". *The Hollywood Reporter*.

[17] Markay, Lachlan (September 28, 2012). "Matt Damon's Anti-Fracking Movie Financed by Oil-Rich Arab Nation". The Heritage Foundation. Retrieved October 1, 2012.

[18] Hargreaves, Steve (October 1, 2012). "Matt Damon fracking film backed by big OPEC member". *CNN Money* (CNN). Retrieved October 1, 2012.

[19] McNary, Dave (August 23, 2012). "Van Sant's 'Promised Land' to arrive Dec. 28". *Variety*.

[20] McClintock, Pamela (December 29, 2012). "Holiday Box Office: 'Django' Narrowly Beats 'Les Mis' on Friday; 'Hobbit' Still No. 1". *The Hollywood Reporter*.

[21] Subers, Ray (December 30, 2012). "Weekend Report: 'Hobbit' Holds Off 'Django' on Final Weekend of 2012". Box Office Mojo. Retrieved December 31, 2012.

[22] Kaufman, Amy (January 6, 2013). "'Texas Chainsaw 3D' is strong No. 1; 'Promised Land' disappoints". *Los Angeles Times*.

[23] "Promised Land (2012)". Box Office Mojo. Retrieved January 9, 2013.

[24] "First Films for the Competition and Berlinale Special". *berlinale*. Retrieved 2012-12-13.

[25] "Prizes of the International Jury". *berlinale*. Retrieved 2013-02-16.

[26] Gettell, Olivia (December 28, 2012). "'Promised Land': Drilling drama lacks depth, critics say". *The Los Angeles Times*.

[27] "Promised Land". Rotten Tomatoes. Retrieved April 21, 2013.

[28] "Promised Land Reviews". Metacritic. Retrieved January 7, 2013.

[29] Scott, A.O. (28 December 2012). "Promised Land with Matt Damon Directed by Gus Van Sant". *New York Times*.

[30] Lacey, Liam (January 4, 2013). "Promised Land: Stellar cast, but the film is a fracking disappointment". Toronto Globe and Mail. Retrieved 7 January 2013.

19.9   External links

- Official website

- *Promised Land* at the Internet Movie Database

- *Promised Land* at Box Office Mojo

- *Promised Land* at Rotten Tomatoes

- *Promised Land* at Metacritic

- *Promised Land* informational page at TakePart

Chapter 20

Stephanie Hallowich, H/W, v. Range Resources Corporation

Stephanie Hallowich and Chris Hallowich, H/W, v. Range Resources Corporation, Williams Gas/Laurel Mountain Midstream, MarkWest Energy Partners, L.P., MarkWest Energy Group, L.L.C., and Pennsylvania Department of Environmental Protection was a case in the Court of Common Pleas of Washington County, Pennsylvania (Civil Division). The trial was held in August 2011.

Chris Hallowich and his wife Stephanie Hallowich operated a farm in Mount Pleasant, Pennsylvania. They sued Range Resources Corporation, Williams Gas/Laurel Mountain Midstream, MarkWest Energy, and the Pennsylvania Department of Environmental Protection for compensation for "health and environmental impacts" from "natural gas development operations".

In the settlement, the plaintiffs received $750,000 from the sale of their home for the purchase of a home elsewhere, but were required to abstain, along with their two minor children, from ever again discussing hydraulic fracturing or Marcellus Shale. The filed Washington County court documents in the case show that the Hallowich family was represented by Peter M. Villari, Esq. and Robert N. Wilkey, Esq. of Villari, Brandes, Kline,P.C., a litigation firm located in Conshohocken, Pennsylvania. The case received significant international notoriety concerning the issue of First Amendment rights as it relates to confidential settlement agreements, gag-orders, and the constitutional rights of minors within the scope of the natural gas drilling industry, operations, and civil litigation. On March 20, 2011, Judge Debbie O'Dell-Senaca, issued an Order and Opinion, reversing a prior trial court's order to seal the settlement dockets, and ordering that the Court record, including the settlement documents be unsealed. Many of the Defendants in the case subsequently appealed, seeking to keep the settlement documents sealed. Eventually, the settlement documents in the Hallowich case were restored and made publicly available.

20.1 External links

- Stephanie Hallowich and Chris Hallowich, H/W, v. Range Resources Corporation; Williams Gas/Laurel Mountain Midstream; MarkWest Energy Partners, L.P.; MarkWest Energy Group, L.L.C.; and Pennsylvania Department of Environmental Protection

- Children Given Gag Order In Pennsylvania Fracking Suit Settlement | Parenting - Yahoo! Shine

- Stephanie Hallowich gas production experiences with Range Resources, MarkWest and Williams

- This 7-Year-Old Is Banned From Talking About Fracking—Ever | Mother Jones

- Children given lifelong ban on talking about fracking | Environment | The Guardian

- Hallowich Unsealed Court Documents, Court of Common Pleas of Washington County, PA, Docket No. 2010-3954

- Can You Silence a Child? Inside the Hallowich Case, Caitlin Dickson, The Daily Beast, UK Guardian, September 1, 2013

Chapter 21

Uses of radioactivity in oil and gas wells

Radioactive sources are used for logging formation parameters. Radioactive tracers, along with the other substances in hydraulic-fracturing fluid, are sometimes used to determine the injection profile and location of fractures created by hydraulic fracturing.[1]

21.1 Use of radioactive sources for logging

Sealed radioactive sources are routinely used in formation evaluation of both hydraulically fractured and non-fracked wells. The sources are lowered into the borehole as part of the well logging tools, and are removed from the borehole before any hydraulic fracturing takes place. Measurement of formation density is made using a sealed caesium-137 source. This bombards the formation with high energy gamma rays. The attenuation of these gamma rays gives an accurate measure of formation density; this has been a standard oilfield tool since 1965. Another source is americium berylium (Am-Be) neutron source used in evaluation of the porosity of the formation, and this has been used since 1950. In a drilling context, these sources are used by trained personnel, and radiation exposure of those personnel is monitored. Usage is covered by licenses from International Atomic Energy Agency (IAEA) guidelines, SU or European Union protocols, and the Environment Agency in the UK. Licenses are required for access, transport, and use of radioactive sources. These sources are very large, and the potential for their use in a 'dirty bomb' means security issues are considered as important. There is no risk to the public, or to water supplies under normal usage. They are transported to a well site in shielded containers, which means exposure to the public is very low, much lower than the background radiation dose in one day.

21.2 Radiotracers and markers

The oil and gas industry in general uses unsealed radioactive solids (powder and granular forms), liquids and gases to investigate or trace the movement of materials. The most common use of these radiotracers is at the well head for the measurement of flow rate for various purposes. A 1995 study found that radioactive tracers were used in over 15% of stimulated oil and gas wells.[2]

Use of these radioactive tracers is strictly controlled. It is recommended that the radiotracer is chosen to have readily detectable radiation, appropriate chemical properties, and a half life and toxicity level that will minimize initial and residual contamination.[3] Operators are to ensure that licensed material will be used, transported, stored, and disposed of in such a way that members of the public will not receive more than 1 mSv (100 mrem) in one year, and the dose in any unrestricted area will not exceed 0.02 mSv (2 mrem) in any one hour. They are required to secure stored licensed material from access, removal, or use by unauthorized personnel and control and maintain constant surveillance of licensed material when in use and not in storage.[4] Federal and state nuclear regulatory agencies keep records of the radionuclides used.[4]

As of 2003 the isotopes Antimony-124, argon-41, cobalt-60, iodine-131, iridium-192, lanthanum-140, manganese-56, scandium-46, sodium-24, silver-110m, technetium-99m, and xenon-133 were most commonly used by the oil and gas industry because they are easily identified and measured.[3][5] Bromine-82, Carbon-14, hydrogen-3, iodine-125 are also used.[3][4]

Examples of amounts used are:[4]

In hydraulic fracturing, plastic pellets coated with Silver-110m or sand labelled with Iridium-192with may be added to a proppant when it is required to evaluate whether a fracturing process has penetrated rocks in the pay zone.[4] Some radioactivity may by brought to the surface at the well head during testing to determine the injection profile and location of fractures. Typically this uses very small (50 kBq) Cobalt-60 sources and dilution factors are such that the activity concentrations will be very low in the topside plant and equipment.[3]

21.3   Regulation in the US

The NRC and approved state agencies regulate the use of injected radionuclides in hydraulic fracturing in the United States.[4]

The US EPA sets radioactivity standards for drinking water.[6] Federal and state regulators do not require sewage treatment plants that accept gas well wastewater to test for radioactivity. In Pennsylvania, where the hydraulic fracturing drilling boom began in 2008, most drinking-water intake plants downstream from those sewage treatment plants have not tested for radioactivity since before 2006.[7] The EPA has asked the Pennsylvania Department of Environmental Protection to require community water systems in certain locations, and centralized wastewater treatment facilities to conduct testing for radionuclides.[8][9][10]

21.4   See also

- List of additives for hydraulic fracturing

- Hydraulic fracturing proppants

21.5   References

[1] Reis, John C. (1976). *Environmental Control in Petroleum Engineering.* Gulf Professional Publishers.

[2] K. Fisher and others, "A comprehensive study of the analysis and economic benefits of radioactive tracer engineered stimulation procedures," *Society of Petroleum Engineers,* Paper 30794-MS, October 1995.

[3] Radiation Protection and the Management of Radioactive Waste in the Oil and Gas Industry (PDF) (Report). International Atomic Energy Agency. 2003. pp. 38–40. Retrieved 20 May 2012. Beta emitters, including ^{3}H and ^{14}C, may be used when it is feasible to use sampling techniques to detect the presence of the radiotracer, or when changes in activity concentration can be used as indicators of the properties of interest in the system. Gamma emitters, such as ^{46}Sc, ^{140}La, ^{56}Mn, ^{24}Na, ^{124}Sb, ^{192}Ir, ^{99}Tc^m, ^{131}I, ^{110}Ag^m, ^{41}Ar and ^{133}Xe are used extensively because of the ease with which they can be identified and measured. ... In order to aid the detection of any spillage of solutions of the 'soft' beta emitters, they are sometimes spiked with a short half-life gamma emitter such as ^{82}Br...

[4] Jack E. Whitten, Steven R. Courtemanche, Andrea R. Jones, Richard E. Penrod, and David B. Fogl (Division of Industrial and Medical Nuclear Safety, Office of Nuclear Material Safety and Safeguards) (June 2000). "Consolidated Guidance About Materials Licenses: Program-Specific Guidance About Well Logging, Tracer, and Field Flood Study Licenses (NUREG-1556, Volume 14)". US Nuclear Regulatory Commission. Retrieved 19 April 2012. labeled Frac Sand...Sc-46, Br-82, Ag-110m, Sb-124, Ir-192

[5] Dina Murphy and Larry Huskins (8 Sep 2006). "letter filed with Department of Environment, New Brunswick, CA" (PDF). Penobsquis, CA government. p. 3. Retrieved 29 July 2012. engineer who works with this radioactive material for a living is exposed to less radiation than an individual who smokes 1.5 packs of cigarettes a day."

[6] US EPA, are EPA's drinking water regulations for radionuclides? What are EPA's drinking water regulations for radionuclides?, accessed 15 Sept. 2013.

[7] "Regulation Lax as Gas Wells' Tainted Water Hits Rivers". New York Times. February 26, 2011.

[8] Urbina, Ian (26 February 2011). "Regulation Lax as Gas Wells' Tainted Water Hits Rivers". *The New York Times*. Retrieved 22 February 2012. The level of radioactivity in the wastewater has sometimes been hundreds or even thousands of times the maximum allowed by the federal standard for drinking water.

[9] Shawn M. Garvin (7 March 2011). "Letter to PADEP re:Marcellus Shale 030711" (PDF). EPA. Retrieved 11 May 2012. ...several sources of data, including reports required by PADEP, indicate that the wastewater resulting from gas drilling operations (including flowback from hydraulic fracturing and other fluids produced from gas production wells) contains variable and sometimes high concentrations of materials that may present a threat to human health and aquatic environment, including radionuclides....Many of these substances are not completely removed by wastewater treatment facilities, and their discharge may cause or contribute to impaired drinking water quality for downstream users, or harm aquatic life...At the same time, it is equally critical to examine the persistence of these substances, including radionuclides, in wastewater effluents and their potential presence in receiving waters.

[10] Ian Urbina (7 March 2011). "E.P.A. Steps Up Scrutiny of Pollution in Pennsylvania Rivers". *The New York Times*. Retrieved 23 February 2012.

Lisburne 1

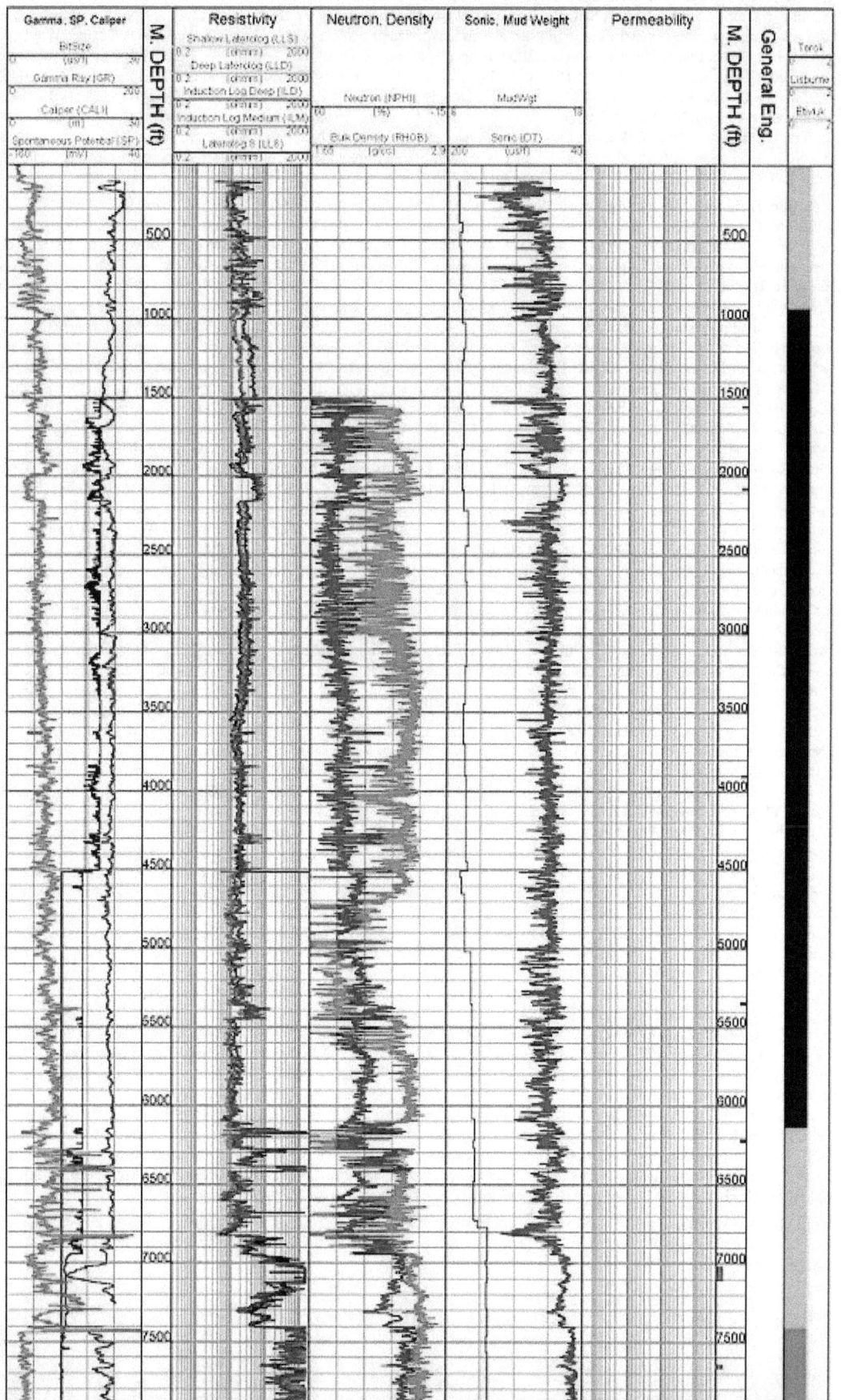

21.6 Text and image sources, contributors, and licenses

21.6.1 Text

- **Environmental impact of hydraulic fracturing in the United States** *Source:* https://en.wikipedia.org/wiki/Environmental_impact_of_ hydraulic_fracturing_in_the_United_States?oldid=688026064 *Contributors:* Dean p foster, Alan Liefting, Rich Farmbrough, Vsmith, Viridi- tas, Drbogdan, Rjwilmsi, Ground Zero, Nihiltres, Kolbasz, Wavelength, Arzel, Welsh, Arthur Rubin, MartinGugino, WilyD, Victorgrigas, Frap, Wen D House, BullRangifer, Dawnseeker2000, Beagel, Shawn in Montreal, DASonnenfeld, Plazak, Swliv, General Epitaph, Imperfect- lyInformed, Unbuttered Parsnip, Mild Bill Hiccup, Jytdog, Dthomsen8, Element16, Download, Sindinero, Jarble, Yobot, AnomieBOT, Citation bot, Camclark04, Gautier lebon, Pinethicket, Jonesey95, Trappist the monk, RjwilmsiBot, Dewritech, GoingBatty, SporkBot, RockMagnetist, Mtndogmedic, ClueBot NG, Smm201`0, Esnascosta, BG19bot, Frze, Mark Arsten, Jakebarrington, BattyBot, Comatmebro, Sa5309, Trichuris trichiura, Kleptopigstar, Faizan, Wuerzele, CJ5Fanatic, Dustin V. S., Veebeejeebee, Hatchingfuture, Trtehp35, Mandelbrony, Monkbot, Koi- jmonop, Paul SciencesPo, Richard elwell, Jdhillinc, Jr703569, Stigs46, Stoney1976, Potato89894, Experienceeducate, Cha.cordy, Oil1236, Hydraulic frac and Anonymous: 28

- **Hydraulic fracturing** *Source:* https://en.wikipedia.org/wiki/Hydraulic_fracturing?oldid=687535819 *Contributors:* Fred Bauder, Egil, Ihcoyc, William M. Connolley, Samw, Hike395, Owen, Lumos3, Bearcat, Rfc1394, Rsduhamel, Alan Liefting, Jeremiah, Ianeiloart, Giddie, Kravietz, Excalibur, GeoGreg, Richardb43, Mike Rosoft, Discospinster, Rich Farmbrough, Kdammers, Vsmith, Smyth, Bender235, Causa sui, Peter Greenwell, Stesmo, Viriditas, La goutte de pluie, Eric Kvaalen, Sjschen, Velella, Dave.Dunford, Kgrr, BlaiseFEgan, RichardWeiss, BD2412, Drbogdan, Rjwilmsi, Jweiss11, Vegaswikian, Mbutts, The wub, Darqcyde, Ground Zero, Kolbasz, Sgartner, Guanxi, DVdm, Bgwhite, Ahpook, Wavelength, Scott Teresi, Jimp, Arzel, DMahalko, Arado, Casey56, Rsrikanth05, Morphh, NawlinWiki, Keithonearth, Abrunete~enwiki, Trovatore, Dhollm, Wknight94, Bdell555, Arthur Rubin, Tomhanna, Dspradau, Wizofaus, Luk, A13ean, MartinGugino, Timeshifter, Kin- tetsubuffalo, Mauls, Gilliam, Ohnoitsjamie, Chris the speller, Jprg1966, Thumperward, Abandon all arr now, Victorgrigas, Deli nk, Solarix, Racklever, VMS Mosaic, Wen D House, BullRangifer, Fuzzypeg, Mwtoews, Salamurai, Captainbeefart, BrownHairedGirl, John, Evenios, Noegenesis, NYCJosh, Peterlewis, IronGargoyle, Pjrm, Majora4, Ricmcc1766, Blackash, Ian peters, CmdrObot, Cartoonmaster, Cooljeanius, MC10, Gogo Dodo, Christian75, Kckid, NorthernThunder, Mglovesfun, Epbr123, Cimbalom, Wikid77, Martin Hogbin, Headbomb, Vertium, Hcobb, Dawnseeker2000, Seaphoto, Blue Tie, SummerPhD, Prolog, Activist, Tillman, Lfstevens, 1Rabid Monkey, Daytona2, Mikenorton, Rothorpe, JamesBWatson, Harelx, Prestonmcconkie, The Anomebot2, Beagel, Sustainableyes, JaGa, VetPsychWars, Mschiffler, Gandydancer, Uriel8, R'n'B, CommonsDelinker, Leyo, Tgeairn, Rlsheehan, Arrivisto, Peter Chastain, Maproom, Shawn in Montreal, RTBoyce, Arms & Hearts, NewEnglandYankee, KylieTastic, Tiggerjay, Equazcion, Reelrt, Vranak, 28bytes, Tourbillon, Kriplozoik~enwiki, BoogaLouie, Lex- ein, Mcewan, Philip Trueman, Tavix, Piperh, Oxfordwang, T0pher17, Aondotri, Wikiisawesome, Onore Baka Sama, Plazak, Meters, Lamro, Antillarum, SylviaStanley, Biscuittin, Swliv, Ketone16, Toddst1, Alexbrn, Rexpilger, Yone Fernandes, William D. Walker, Msiner, FghI- Jklm, Capitalismojo, Denisarona, A21sauce, Mrfebruary, NickCT, General Epitaph, ImperfectlyInformed, Unbuttered Parsnip, Der Golem, JTSchreiber, Niceguyedc, Deselliers, Kitsunegami, Excirial, Crywalt, Moreau1, Sun Creator, Arjayay, SchreiberBike, The Yowser, XLinkBot, Jytdog, Gerhardvalentin, Avoided, Rreagan007, WikHead, Addbot, Mneuner, PetroleumAge, MrOllie, Download, CarsracBot, RTG, Favo- nian, Sindinero, Zorrobot, Jarble, Prawlings, Ben Ben, Legobot, Yobot, EdwardLane, Fraggle81, Librsh, AnomieBOT, DemocraticLuntz, Yukonpeegs, Archon 2488, Jim1138, Jo3sampl, Materialscientist, Ckruschke, Citation bot, Quebec99, LilHelpa, Srich32977, GrouchoBot, Owenh000, SassoBot, Pikachu sensei, GESICC, Celuici, FrescoBot, Pnin2006, Blackguard SF, Shaibalahmar, Citation bot 1, Chenopodi- aceous, Casprings, IrrtNie, Gautier lebon, Leosaraceni, Pinethicket, Bibliophile227, Jonesey95, Andynct, Rushbugled13, Jschnur, RedBot, CanberraBulldog, Serols, Elekhh, FoxBot, Trappist the monk, Jordgette, AHeneen, S1id3r0, GregKaye, Dinamik-bot, Vrenator, Jeffrd10, Mean as custard, RjwilmsiBot, J36miles, EmausBot, John of Reading, Soporyc, WikitanvirBot, Mabuzi, Jdkag, Laurabauer, GoingBatty, RA0808, Gimmetoo, Wikipelli, K6ka, ZéroBot, John Cline, Josve05a, Geperdo, Compdude123, Agent0060, Druzhnik, Xen1977, Ire2500, Rcsprinter123, Joshua Doubek, IGeMiNix, Rostz, Braculus, Donner60, Bulwersator, RockMagnetist, Sonicyouth86, ClueBot NG, Tonystew- art14, Somedifferentstuff, Gilderien, Satellizer, Iloveandrea, Joefromrandb, Coastwise, Smm201`0, WebMaven2000, Snotbot, AdamRBrooks, Kevin Gorman, Law of Entropy, Rcorym15, Joel B. Lewis, Widr, Esnascosta, Helpful Pixie Bot, HMSSolent, Strike Eagle, Candleabra- cadabra, Wbm1058, Bibcode Bot, Denovoid, JohnEdit21, BG19bot, NewsAndEventsGuy, Yendor of yinn, Teach267, The Mark of the Beast, IlllIIIlOIllllIII, Neøn, OpenMind, MusikAnimal, Frze, Amp71, Bonnie13J, Stephenwanjau, Mark Arsten, IraChesterfield, StephanieF79, Mis- misanthrope, Cameron6426a, Terrance.cunningham, Maxellus, EnviroE, Gorthian, Joydeep, T.hetton, Chase.alton3, KelseyBrook, FertileCat- fish, OChemie, Carbon13neutral, Polmandc, Fracksand, Thegreatgrabber, Klilidiplomus, Hoppenhe, EnvPolKAL, Adamuwt2011, Proxyma, 17mansure, BrianWo, BattyBot, Easshale, EnergeticsAnalyst, DeRanged Squirrel, Emerituseditor, Palpbert, Jfsparks, Brandiwessel, Nfara- guna, MissLoveIsMyWeapon, T.Dooshswag, Brentonchina, Sophk, The Illusive Man, Timothy Gu, ChrisGualtieri, GrizBizzy, Roche398, Aldenallen1, Odewey, AJKeown, Kyle assassinz, Christine1223, EuroCarGT, Joseph.ruggiero, Illia Connell, Runtzz, Lhcollins, IjonTichyIjon- Tichy, Zstallman, Writer1502, EagerToddler39, JurgenNL, SoledadKabocha, Mogism, Cerabot~enwiki, Conor Strong, CuriousMind01, Frac- master, Raiderredjd, Numbermaniac, Scgtcheck, Proper Stranger, Kleptopigstar, Newmanmu, Jpmsd, Mdicato, Project Osprey, JustAMuggle, Elstree-230, Lgfcd, Gabby Merger, Whitepanthera, Exenola, Faizan, Sharmeka-winnsboro, Men404, PlanetEditor, Lbriggs, Tothesungod, Nogginquest, PhantomTech, Jeffthajamaican, Jabx2, Wuerzele, Amanda.caggiano, CJ5Fanatic, Cjpepino, EllenCT, DavidLeighEllis, Niko- lai778, Exesop, Luchi1223, Ellisnyc, ReconditeRodent, Alanis435, Firebrandcentral, Lil wayne666, GypsyEyes, 1822002hadrian, Hadrian0, Roman26, Mandruss, Hays452, Mutantkarma, Caswivel, Noyster, Top5percent, Lpfleischmann, Sr75080, Chris.hooper78, Kikishiki, Patriv- elaa, Maxluke, Batuhanguven94, G S Palmer, 7Sidz, Ur mumcxhbgvhjgvasdjk, Salubrious Toxin, Wiki at Royal Society John, Monkbot, Renamed user 51g7z61hz5af2azs6k6, Leegrc, DoctorBrominestein, Vieque, Cochandl77, Lwiklund, Paul SciencesPo, Kennywpara, Richard elwell, Savannahb022, Scarlettail, Sygac, Nikify, Hamzahno, Narddawg, Jr703569, El Kaput, Awesomeburner, Umkan, Njingo, Aimrux, Ck- err65, Richard Yin, HACNY, Cranberry Products, Sourgosling, A guy saved by Jesus, Mundopopular, HIyer01, Donnywinston, Griffindell, TFFX, Horses photos, Berg3usa, Narky Blert, Rhod08, Fatetrevor, Crystallizedcarbon, Beanstash, Pcook22, Antigomer, Tymon.r, Stoney1976, Turtle Guillotine, Thecrew1021, Xdspellmaster, Hvaara, Tjhlax, Mountainlove, 234 family, Catherath, Emnags, Dsc92, Rmbenson, Argo-e, KasparBot, Lloydwas jones123, Longauria, CopperPhoenix, JamesWinston1 and Anonymous: 538

- **Regulation of hydraulic fracturing** *Source:* https://en.wikipedia.org/wiki/Regulation_of_hydraulic_fracturing?oldid=672070683 *Contrib- utors:* Drbogdan, Wavelength, Thumperward, Dawnseeker2000, DASonnenfeld, AlbinoFerret, AnomieBOT, RockMagnetist, OccultZone, 7Sidz and Anonymous: 1

- **List of additives for hydraulic fracturing** *Source:* https://en.wikipedia.org/wiki/List_of_additives_for_hydraulic_fracturing?oldid=643585826 *Contributors:* Alan Liefting, Ekem, Edgar181, Smokefoot, Cydebot, Martin Hogbin, Katalaveno, Auntof6, Jytdog, RockMagnetist, Smm201`0, T.hetton, BattyBot and Anonymous: 6

- **Baldwin Hills Dam disaster** *Source:* https://en.wikipedia.org/wiki/Baldwin_Hills_Dam_disaster?oldid=672533088 *Contributors:* SimonP, Docu, IceKarma, Wetman, Cyrius, Alobodig, Eric Shalov, Larry Grossman, Miwasatoshi, Emerson7, Rjwilmsi, Vegaswikian, SpuriousQ, Epipelagic, Hmains, Will Beback, Cydebot, Dawnseeker2000, Minnaert, Aeh4543, Jllm06, Cgingold, DandyDan2007, Zuejay, Rocket71048576, Imasleepviking, Hinderkinder, Trackinfo, Steve72, Tanvir Ahmmed, Umbertod, Stepheng3, XLinkBot, Manycars, Kbdankbot, Addbot, ProfessorXY, Lightbot, Captain Quirk, Imveracious, Griffinofwales, Citation bot 1, Trappist the monk, RjwilmsiBot, Androstachys, John of Reading, Look2See1, Shannon1, H3llBot, Helpful Pixie Bot, BG19bot, Jackswelters, Comfr, BattyBot, JustAMuggle, Monkbot, Stanislaw.stawowy, AlexBing60 and Anonymous: 16

- **Canadian Association of Petroleum Producers** *Source:* https://en.wikipedia.org/wiki/Canadian_Association_of_Petroleum_Producers?oldid=674972930 *Contributors:* Bearcat, Bobblewik, JamesTeterenko, PaulHanson, Woohookitty, Zippo, DESiegel, Bgwhite, Ikar.us, Malcolma, Nick Dillinger, Cavenba, Chris the speller, Ottawakismet, Flipperinu, CmdrObot, Cydebot, Alaibot, Pmbcomm, Oceanflynn, InvestInCanada, Pjoef, Carpasian, Unbuttered Parsnip, 718 Bot, XLinkBot, Dthomsen8, Yobot, Nikonoff, AnomieBOT, LilHelpa, Moxy, Kibi78704, RjwilmsiBot, John of Reading, Cappweb, Gilderien, Newyorkadam, Hackofalltrades, JLangberg and Anonymous: 13

- **Canol shale play** *Source:* https://en.wikipedia.org/wiki/Canol_shale_play?oldid=675036245 *Contributors:* Bearcat, Geo Swan, Ground Zero, Gbawden, Alvin Seville and BattyBot

- **Chevron CRUSH** *Source:* https://en.wikipedia.org/wiki/Chevron_CRUSH?oldid=624567091 *Contributors:* Alan Liefting, H Padleckas, Beagel, Lamro, Lightmouse, WikHead, JimVC3, Citation bot 1, RjwilmsiBot, Helpful Pixie Bot and Anonymous: 2

- **Environmental impact of hydraulic fracturing** *Source:* https://en.wikipedia.org/wiki/Environmental_impact_of_hydraulic_fracturing?oldid=686813474 *Contributors:* Fred Bauder, Markhurd, BD2412, Rjwilmsi, Welsh, Bdell555, Gilliam, Victorgrigas, BullRangifer, Iridescent, Blackash, CmdrObot, Martin Hogbin, JustAGal, Nick Number, Tillman, Beagel, Oceanflynn, Ontarioboy, DASonnenfeld, Plazak, Lamro, Alexbrn, Megiddo1013, Jytdog, Dthomsen8, Jarble, Yobot, AnomieBOT, Archon 2488, Quebec99, Shadowjams, Gautier lebon, Jonesey95, Trappist the monk, RjwilmsiBot, GoingBatty, Midas02, RockMagnetist, ClueBot NG, Smm201`0, BG19bot, BattyBot, Kupiakos, Sa5309, Viewmont Viking, Frosty, Kleptopigstar, Rasforte, JustAMuggle, VanishedUser 2313214sad1, Wuerzele, Dustin V. S., EllenCT, Mandruss, Ginsuloft, Chris.hooper78, Robevans123, Monkbot, Paul SciencesPo, Kennywpara, BethNaught, BeecherP, F.Nonsense, Lfrankbalm, Stoney1976, Abierma3, Prodigy73, Oil1236 and Anonymous: 26

- **Exemptions for hydraulic fracturing under United States federal law** *Source:* https://en.wikipedia.org/wiki/Exemptions_for_hydraulic_fracturing_under_United_States_federal_law?oldid=685496995 *Contributors:* Dcoetzee, Bearcat, Stuartyeates, Ground Zero, Woodshed, CmdrObot, Martin Hogbin, Keith D, Plazak, Calliopejen1, Yobot, AnomieBOT, Eumolpo, RockMagnetist, Snotbot, Kevin Gorman, Aarf613, BG19bot, Parathin81, Cjpepino, Totranm, StacyPF, Monkbot, Aimrux and Anonymous: 4

- **ExxonMobil Electrofrac** *Source:* https://en.wikipedia.org/wiki/ExxonMobil_Electrofrac?oldid=617330645 *Contributors:* Alan Liefting, Pascal666, Patken4, Beagel, Citation bot 1, RjwilmsiBot, Jdkag, Helpful Pixie Bot, Monkbot and Anonymous: 2

- **Fracking hose** *Source:* https://en.wikipedia.org/wiki/Fracking_hose?oldid=644174098 *Contributors:* Chris the speller, Anthony Bradbury, GrahamHardy, Biscuittin, Mrt3366, ChrisGualtieri, Fox2k11, Altered Walter, Firehosemart, Kenchak91 and Zeshan15

- **The FracTracker Alliance** *Source:* https://en.wikipedia.org/wiki/The_FracTracker_Alliance?oldid=643123509 *Contributors:* GrahamHardy, GoingBatty, Kevin Gorman, EricEnfermero and Anonymous: 1

- **Fracturing Responsibility and Awareness of Chemicals Act** *Source:* https://en.wikipedia.org/wiki/Fracturing_Responsibility_and_Awareness_of_Chemicals_Act?oldid=594132168 *Contributors:* Plandu, Alan Liefting, Canterbury Tail, Viriditas, Landroni, Rjwilmsi, Ground Zero, Arzel, Ccgrimm, Wainstead, SmackBot, Ohnoitsjamie, Reywas92, Zigzig20s, Brownout, Beagel, XLinkBot, Addbot, Tassedethe, Sindinero, Luckas-bot, AnomieBOT, Clearvilletimes, John of Reading, Gheadgordon, Rostz, ClueBot NG, Smm201`0, BG19bot, Easshale and Anonymous: 7

- **Hydraulic fracturing proppants** *Source:* https://en.wikipedia.org/wiki/Hydraulic_fracturing_proppants?oldid=686395880 *Contributors:* Vsmith, Giraffedata, Kolbasz, Ahpook, A5b, Dl2000, Lamiot, Ztolstoy, Martin Hogbin, Mikenorton, Beagel, Cathwoodul, Westfalr3, Malcolmxl5, Jytdog, Wikiuser100, Addbot, Elemented9, Yobot, AnomieBOT, HRoestBot, Jonesey95, RoadTrain, RjwilmsiBot, RockMagnetist, ClueBot NG, Smm201`0, Helpful Pixie Bot, DeRanged Resources, Nothing gold can stay, ChrisGualtieri, Mediran, Blairwal, Monkbot, Kennywpara, Jbsingh86, Pcook22, Stoney1976 and Anonymous: 17

- **Hydro-slotted perforation** *Source:* https://en.wikipedia.org/wiki/Hydro-slotted_perforation?oldid=654798594 *Contributors:* Bearcat, Wavelength, Magioladitis, Katharineamy, WOSlinker, Lamro, Cindamuse, Addbot, Greyhood, Yobot, SwisterTwister, AnomieBOT, FrescoBot, WQUlrich, SBaker43, Mrt3366, Solixcanada, Ufuoma arhagba and Anonymous: 1

- **Mohamed Yousef Soliman** *Source:* https://en.wikipedia.org/wiki/Mohamed_Yousef_Soliman?oldid=660979698 *Contributors:* Rubaisport, Waacstats, DGG, Yobot, Cncmaster, BattyBot, Mogism, Aloneinthewild, Eliaspirayesh and Anonymous: 4

- **Oil and Gas Commission** *Source:* https://en.wikipedia.org/wiki/Oil_and_Gas_Commission?oldid=647043229 *Contributors:* Plasma east, SmackBot, LLP, Colonies Chris, Whpq, Cydebot, Biruitorul, Rich257, Keefer4, Yobot, Daniele Pugliesi, DrilBot, Mean as custard, Helpful Pixie Bot, Chelsea.green, DoctorKubla, Wuerzele and Anonymous: 4

- **Promised Land (2012 film)** *Source:* https://en.wikipedia.org/wiki/Promised_Land_(2012_film)?oldid=683735421 *Contributors:* Btphelps, Tonymaric, JustPhil, Erik, BDD, Noclador, Marketdiamond, Arthur Rubin, Emurphy42, Evensong, Luigibob, Morganfitzp, Lugnuts, Paranoidgoat, Freddiem, Guy Macon, Varnent, Bovineboy2008, CrossoverManiac, Macae, JukoFF, Flyer22 Reborn, Goustien, Jrobb525, Patrick Rogel, Trivialist, Cliff1911, Addbot, Nickthedangerous, HannibalV, Yobot, AnomieBOT, ChocolateBlender, Cresix, Eugene-elgato, Sock, Ratim, HRoestBot, Arbero, Jedi94, Tofutwitch11, Tbhotch, Stephenmeb, AsceticRose, Marek Koudelka, Lacon432, InfamousPrince, Rusted AutoParts, Crakkerjakk, Gareth Griffith-Jones, DrBeWa, Amymc12, BattyBot, EdenCole, ChrisGualtieri, Mstarr3, Kanghuitari, Dexbot, ExclusiveAgent, Eric 059, BTRand1, ReutersTV, OilAndGasProperties, Rockiesoil, Caffeinated42, Tina Gerhardt, HunterLeeLogan, DavidLeighEllis, Qc1okay, RRRFootprint, TropicAces and Anonymous: 68

- **Stephanie Hallowich, H/W, v. Range Resources Corporation** *Source:* https://en.wikipedia.org/wiki/Stephanie_Hallowich%2C_H/W%2C_v._Range_Resources_Corporation?oldid=664629974 *Contributors:* Alan Liefting, Wavelength, Jhawkinson, Good Olfactory, Lightlowemon, Dewritech, GoingBatty and Writingwitwords

- **Uses of radioactivity in oil and gas wells** *Source:* https://en.wikipedia.org/wiki/Uses_of_radioactivity_in_oil_and_gas_wells?oldid=684406558 *Contributors:* Edward, BarkingFish, Rjwilmsi, Vegaswikian, Kolbasz, Bgwhite, Wavelength, Arzel, Arthur Rubin, Martin Hogbin, Mikenorton, Beagel, Plazak, Jytdog, Yobot, Guy1890, AnomieBOT, FrescoBot, John of Reading, Djembayz, RockMagnetist, Smm201`0, Snotbot, BG19bot, Frze, Mark Arsten, Lazord00d, BattyBot, EagerToddler39, Mandruss, Monkbot, Kennywpara, Jr703569 and Anonymous: 3

21.6.2 Images

- **File:2011-2014_water_use_for_fracking.jpg** *Source:* https://upload.wikimedia.org/wikipedia/commons/e/e9/2011-2014_water_use_for_fracking.jpg *License:* Public domain *Contributors:* http://www.usgs.gov/newsroom/images/2015_06_30/water_use_for_fracking.jpg *Original artist:* USGS

- **File:Aegopodium_podagraria1_ies.jpg** *Source:* https://upload.wikimedia.org/wikipedia/commons/b/bf/Aegopodium_podagraria1_ies.jpg *License:* CC-BY-SA-3.0 *Contributors:* Own work *Original artist:* Frank Vincentz

- **File:Ambox_globe_content.svg** *Source:* https://upload.wikimedia.org/wikipedia/commons/b/bd/Ambox_globe_content.svg *License:* Public domain *Contributors:* Own work, using File:Information icon3.svg and File:Earth clip art.svg *Original artist:* penubag

- **File:Ambox_important.svg** *Source:* https://upload.wikimedia.org/wikipedia/commons/b/b4/Ambox_important.svg *License:* Public domain *Contributors:* Own work, based off of Image:Ambox scales.svg *Original artist:* Dsmurat (talk · contribs)

- **File:BarnettShaleDrilling-9323.jpg** *Source:* https://upload.wikimedia.org/wikipedia/commons/5/5d/BarnettShaleDrilling-9323.jpg *License:* CC BY-SA 3.0 *Contributors:* Own work *Original artist:* Loadmaster (David R. Tribble)

- **File:Bhdwrview1963.jpg** *Source:* https://upload.wikimedia.org/wikipedia/commons/9/94/Bhdwrview1963.jpg *License:* Public domain *Contributors:* screenshot dwr report on failure *Original artist:* dept water resources

- **File:Canadian_Association_of_Petroleum_Producers_logo.png** *Source:* https://upload.wikimedia.org/wikipedia/en/7/7b/Canadian_Association_of_Petroleum_Producers_logo.png *License:* Fair use *Contributors:* From files provided by CAPP; specified in their corporate branding standards *Original artist:* ?

- **File:Chevron_Oil_Shale_Project.PNG** *Source:* https://upload.wikimedia.org/wikipedia/en/9/9d/Chevron_Oil_Shale_Project.PNG *License:* PD *Contributors:*

  From the US Bureau of land management Environmental Assessment on Chevron Oil Shale Project in Colorado - http://www.co.blm.gov/wrra/documents/Figures.pdf *Original artist:* ?

- **File:Commons-logo.svg** *Source:* https://upload.wikimedia.org/wikipedia/en/4/4a/Commons-logo.svg *License:* ? *Contributors:* ? *Original artist:* ?

- **File:Crystal_energy.svg** *Source:* https://upload.wikimedia.org/wikipedia/commons/1/14/Crystal_energy.svg *License:* LGPL *Contributors:* Own work conversion of Image:Crystal_128_energy.png *Original artist:* Dhatfield

- **File:Edit-clear.svg** *Source:* https://upload.wikimedia.org/wikipedia/en/f/f2/Edit-clear.svg *License:* Public domain *Contributors:* The *Tango! Desktop Project. Original artist:*

  The people from the Tango! project. And according to the meta-data in the file, specifically: "Andreas Nilsson, and Jakub Steiner (although minimally)."

- **File:Flag_of_British_Columbia.svg** *Source:* https://upload.wikimedia.org/wikipedia/commons/b/b8/Flag_of_British_Columbia.svg *License:* Public domain *Contributors:* Province of British Columbia [2] *Original artist:* Original concepts of Arthur John Beanlands; Ambrose Lee, York Herald (1906); and Conrad Swan, York Herald (1987), College of Arms, London.[3]. Rendered into SVG by -xfi-.

- **File:Frac_job_in_process.JPG** *Source:* https://upload.wikimedia.org/wikipedia/commons/9/90/Frac_job_in_process.JPG *License:* CC BY-SA 3.0 *Contributors:* Own work *Original artist:* Joshua Doubek

- **File:Halliburton_Frack_Job_in_the_Bakken.JPG** *Source:* https://upload.wikimedia.org/wikipedia/commons/1/11/Halliburton_Frack_Job_in_the_Bakken.JPG *License:* CC BY-SA 3.0 *Contributors:* Own work *Original artist:* Joshua Doubek

- **File:Hydraulic_Fracturing-Related_Activities.jpg** *Source:* https://upload.wikimedia.org/wikipedia/commons/7/73/Hydraulic_Fracturing-Related_Activities.jpg *License:* Public domain *Contributors:* http://www2.epa.gov/hfstudy/hydraulic-fracturing-water-cycle *Original artist:* US Environmental Protection Agency

- **File:HydroFrac.png** *Source:* https://upload.wikimedia.org/wikipedia/commons/7/75/HydroFrac.png *License:* CC BY-SA 3.0 *Contributors:* Own work *Original artist:* Mikenorton

- **File:HydroFrac2.svg** *Source:* https://upload.wikimedia.org/wikipedia/commons/a/ad/HydroFrac2.svg *License:* CC BY-SA 3.0 *Contributors:* Own work *Original artist:* Mikenorton

- **File:LI1LOG.jpg** *Source:* https://upload.wikimedia.org/wikipedia/commons/5/5e/LI1LOG.jpg *License:* Public domain *Contributors:* http://energy.cr.usgs.gov/OF00-200/WELLS/LISBURN1/LAS/LI1LOG.JPG *Original artist:* USGS

- **File:Process_of_mixing_water_with_fracking_fluids_to_be_injected_into_the_ground.JPG** *Source:* https://upload.wikimedia.org/wikipedia/commons/9/95/Process_of_mixing_water_with_fracking_fluids_to_be_injected_into_the_ground.JPG *License:* CC BY-SA 3.0 *Contributors:* Own work *Original artist:* Joshua Doubek

- **File:Promised_Land_Poster_(2012).jpg** *Source:* https://upload.wikimedia.org/wikipedia/en/f/fd/Promised_Land_Poster_%282012%29.jpg *License:* Fair use *Contributors:*

  May be found at the following website: IMP Awards *Original artist:* ?

21.6.3 Content license

www.ingramcontent.com/pod-product-compliance
Lightning Source LLC
Chambersburg PA
CBHW080641180526
45168CB00008B/3256